IEE CONTROL ENGINEERIN

SERIES EDITORS: G.A. MONTGOM
PROF. H. NICHO

Transducers in digital systems

Volumes in this series:

Volume 1 Multivariable control theory
J.M. Layton
Volume 2 Lift traffic analysis, design and control
G.C. Barney and S.M. dos Santos
Volume 3 Transducers in digital systems
G.A. Woolvet
Volume 4 Supervisory remote control systems
R.E. Young
Volume 5 Structure of interconnected systems
H. Nicholson
Volume 6 Power system control
M.J.H. Sterling
Volume 7 Feedback and multivariable systems
D.H. Owens
Volume 8 A history of control engineering:
1800–1930
S. Bennett

IEE CONTROL ENGINEERING SERIES 3

Transducers in digital systems

G.A. WOOLVET, M.Tech., D.I.C., C.Eng., M.I.Mech.E., M.R.Ae.S., M.I.E.E., M.Inst.M.C.

Head of School of Mechanical, Aeronautical and Production Engineering

Kingston Polytechnic
Kingston upon Thames
Surrey

PETER PEREGRINUS LTD.
on behalf of the
Institution of Electrical Engineers

Published by: The Institution of Electrical Engineers, London
and New York
Peter Peregrinus Ltd., Stevenage, UK, and New York

First published 1977
© 1977: Institution of Electrical Engineers
Reprinted as a paperback 1979

All rights reserved. No part of this publication may be reproduced, stored in a retrieval system, or transmitted in any form or by any means—electronic, mechanical, photocopying, recording or otherwise—without the prior written permission of the publisher

ISBN: 0 906048 13 3

Printed in Great Britain by A. Wheaton & Co. Ltd., Exeter

Contents

	Page
Preface	viii

1 Digital systems — 1
   1.1 Introduction, 1
   1.2 Data loggers, 1
   1.3 Computer-controlled systems, 3
       Analog systems and multiplexers, 7
       Highway systems, 9
       Analog and digital transducers, 10
   1.4 Reference, 12

2 Angular digital encoders — 13
   2.1 Absolute encoders, 14
       Contact encoders, 14
       Scan problems, 16
       U-scan, 20
       Magnetic shaft encoder, 21
       Optical encoders, 24
       Amplifiers, 26
       Additional facilities, 28
       Optical resolver, 31
   2.2 Incremental shaft encoders, 36
       Position logic, 37
       Use of synchros as incremental encoders, 40
   2.3 Digital tachometers, 44
       Electromagnetic pulse tachometers, 47
       Capacitive tachometers, 49
   2.4 References, 50

3 Frequency dependent transducers — 51
   3.1 Voltage/frequency converters, 51
       Measurement techniques, 54
   3.2 Transducer oscillators, 55
       Thermistor temperature-to-frequency converter, 55
       Quartz temperature/frequency converter, 57
       Quartz pressure/frequency converter, 58
       Vibrating string and vibrating beam transducers, 59
       Vibrating diaphragm pressure transducer, 63

　　　　3.3　Vibrating cylinder transducers, 64
　　　　　　　Gas pressure transducer, 66
　　　　　　　Gas density transducer, 70
　　　　　　　Mass flow measurement, 73
　　　　　　　Liquid density transducer, 75
　　　　3.4　References, 77
4　**Digital linear transducers**　　　　　　　　　　　　　　78
　　　　4.1　Using rotary encoders, 78
　　　　4.2　Linear encoders, 80
　　　　　　　Incremental encoder, 83
　　　　　　　Mechanical details, 85
　　　　　　　Inductosyns, 86
　　　　4.3　Moiré fringe techniques, 87
　　　　　　　Increasing the resolution, 89
　　　　　　　Absolute transducer using grating, 92
　　　　4.4　References, 96

5　**Analog conversion methods**　　　　　　　　　　　　　97
　　　　5.1　Installation, 98
　　　　5.2　Multiplexers, 98
　　　　　　　Mechanical scanners, 99
　　　　　　　Multiplexers using relays, 101
　　　　　　　Transistor multiplexers, 103
　　　　　　　Noise from multiplexers, 103
　　　　5.3　Signal conditioning, 104
　　　　　　　Low-pass filter, 105
　　　　　　　Position of filters, 108
　　　　　　　Amplifiers, 109
　　　　　　　Signal filtering, 109
　　　　　　　Digital filtering, 112
　　　　5.4　Analog-to-digital converters, 113
　　　　　　　Aperture time, 114
　　　　　　　Accuracy of conversion, 114
　　　　　　　Sources of error, 115
　　　　　　　Noise error, 116
　　　　　　　Nonlinearity error, 116
　　　　　　　Gain error, 116
　　　　　　　Offset error, 118
　　　　　　　Conversion techniques, 118
　　　　　　　Analog methods, 118
　　　　　　　Feedback methods, 121
　　　　　　　Digital ramp, 124
　　　　　　　Reversible counter, 125
　　　　　　　Successive approximation, 125
　　　　　　　Conclusion, 126
　　　　5.5　References, 127

6　**Synchro/resolver conversion**　　　　　　　　　　　　129
　　　　6.1　Synchro systems, 129
　　　　　　　Synchro pair, 129
　　　　　　　Control differential, 132
　　　　　　　Resolver, 133

Contents   vii

            Torque units, 133
            Brushless synchros, 134
            Two-speed synchro system, 135
    6.2   Tracking converters, 136
            Resolver format, 136
            Phase-shift converters, 137
            Function-generator converters, 140
            Tracking converters, 144
            High-resolution tracking converters, 148
    6.3   Sampling converters, 151
            Sampling techniques, 153
            Successive-approximation converters, 154
            Harmonic-oscillator converter, 155
            Converter errors, 159
            Digital-to-synchro conversion, 162

7   Other techniques   163
    7.1   Digital position sensors, 163
            Inductance sensors, 165
            Capacitance sensors, 167
    7.2   Force balance feedback transducers, 168
    7.3   Magnetic transducers, 170
            Magnetic matrix transducer, 171
            Magnetic recording transducer, 172
    7.4   Radiation transducers, 173
    7.5   Vortex transducers, 176
    7.6   References, 177

Appendix 1 Binary codes   178
Appendix 2 Analog transducers   184
Index   191

Preface

This text provides a survey of transducers that give a digital output. These transducers are mainly used in computer-based systems or where 'digital' readout is required. While two-stage digital logic seems an obvious and natural way to transmit and manipulate data, there do not appear to be any natural phenomena that give a digitally coded output which varies in response to, say, force, flow rate, displacement or temperature. The only possible exceptions are, perhaps, those devices that provide an output which varies in frequency, in response to some change in a physical parameter. Devices of this nature are described. Any other digital transducer generally requires conversion to provide a digital output from a primary analog signal.

The increasing use of digital systems for measurement and control has led to a growing interest in digital transducers, and many special techniques and devices have been developed, some being in current use. This text sets out to describe some of these methods and to survey other interesting developments that are not yet commercial propositions.

The book should be useful to engineers working with on-line computer systems and for graduates and postgraduate students concerned with instrumentation systems.

Chapter 1

Digital systems

1.1 Introduction

The growth of digital systems, reflected in the application of digital computers and the requirements for digital readouts for instrument systems and transducers, has created an interest in digital transducers and in the interfacing of analog transducers with digital systems. The advent of minicomputers and microprocessors now makes it economically acceptable for small manufacturing processes and data-collection and recording systems to be computer controlled. The decreasing costs of mini- and microprocessors make it an important matter to keep the costs of the associated transducers and their interface equipment to a minimum. It is in these situations that transducers with digital outputs play an important part.

Unfortunately, there are only a few direct digital transducer techniques, but there are a number of adaptations of analog transducers and analog-to-digital (A/D) converters to provide digital outputs from some measured parameter. Some of these methods are described in subsequent chapters. A brief review of analog transducers is given in Appendix 2. The remainder of this chapter describes briefly some of those digital systems within which transducers and A/D converters have to operate.

1.2 Data loggers

Most data loggers are designed to receive signals from analog transducers of the types described in Appendix 2. The function of a logger is to access the outputs of a number of transducers in sequence, and to convert the measurement into a form that can be recorded, either on punched paper tape or in typed form, often using a teletype for both purposes. Other recording methods are also used, e.g. magnetic tapes, often in the form of small cassettes.

Most commercial data loggers make provision for a variety of different inputs, with particular emphasis on the low-level signals derived from resistance strain gauges and thermocouples. Any other transducer output can, of course, be handled, but the input must be a d.c. signal. That is, the value of the signal, usually a voltage, must represent the measure of the parameter being measured at the instance that it is sampled by the data logger.

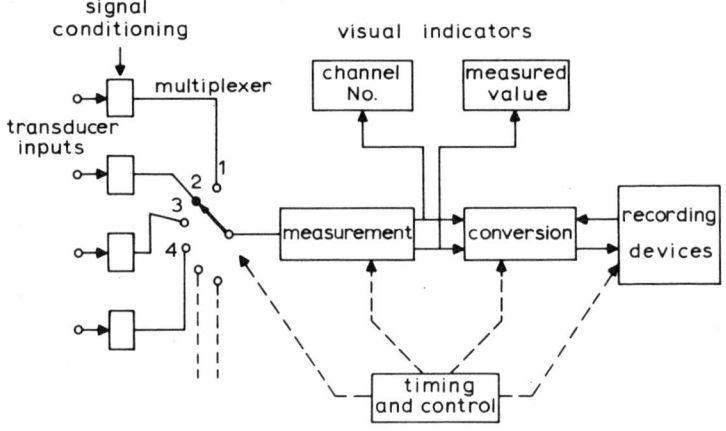

Fig. 1.1   Basic data logger

The essential features of a data logger (Fig. 1.1) are:

(i)   Signal conditioning circuits: these may be made up of the power supplies for the bridge networks in the transducers, as well as the circuits to 'clean up' the input signals and scale them for the rest of the system;
(ii)   Multiplexer: this is a switching arrangement, either electromagnetic or electronic, in which each channel is sampled in turn;
(iii)   Measurement circuit: this monitors the magnitude of the input signal, and outputs other signals to identify the channel on a display, and also the magnitude (and sign) of the measured parameter.
(iv)   Converter: the analog voltage is not in a convenient form for use in punch-tape or teletype systems, so that the converter converts the analog signal to a digital signal through an A/D converter. The signal is

usually in the form of a parallel logic, in straight binary or binary-coded decimal. It is then necessary to feed the recording device in serial form.

(v)   Timing and control: it is necessary to control the rate of sampling, and a central oscillator is used as a timing system (this can also be used as a clock to record the time of day (and the day and month) to the recording device). The maximum speed of sampling is generally limited by the speed of the recording device, and here magnetic tape recorders can offer much higher speeds than other types. The control circuits can switch the system incrementally through the input channels (or this can be achieved manually) or can control single or repetitive cycles. It is necessary to ensure that ample time is given for the data (and the channel number) to be converted to digital signals and for these signals to be translated by the recording device into the required form (it usually types or records a series of separate digits representing each piece of data or channel number).

Other information, to indicate spaces between the channel number and data and carriage-return/line-feed signals, for the teletype, and similar separating signals for punched tape or magnetic tape, must be encoded, and all this must be achieved before the next channel is sampled. It may also be necessary to ensure that changes in the measured parameter do not in the meantime cause the data to the recorder to change. This can be avoided by a sample-and-hold circuit incorporated in the measurement block.

1.3   Computer-controlled systems

Computer-controlled industrial processes have been developed (from the systems usually controlled by analog controllers) which generate control signals in response to input demands compared to the measured values, as determined by a transducer. A variety of different types of transducer are used, e.g. potentiometers, strain gauges, LDVTs, or synchros. In the final stages, the parameter being measured, e.g. position, temperature, pressure, flow, velocity, or density, is defined by an electrical voltage (although some systems employ controllers in which the signals from the transducers are currents).

The limitations of pure analog systems lie in the limited processing of data that can be accomplished, and in the difficulties of changing the type of data manipulation if this is found to be necessary. Additional costs are incurred if it is necessary to provide digital readouts of any analog signals, since it is necessary to provide an A/D converter for each such readout.

4   Digital systems

The introduction of a digital computer to a system permits the range of signal manipulation to be considerably extended, giving, in some cases, a quick analysis to determine characteristics that cannot be measured directly. It also provides an easy method of processing a control signal. A further advantage is that the computer can condition the electrical voltages representing the measurements to identify any rogue readings and/or filter any unwanted 'noise' present.

A typical control system using a digital computer is shown in Fig. 1.2. This illustrates conventional on-line digital control, in which analog

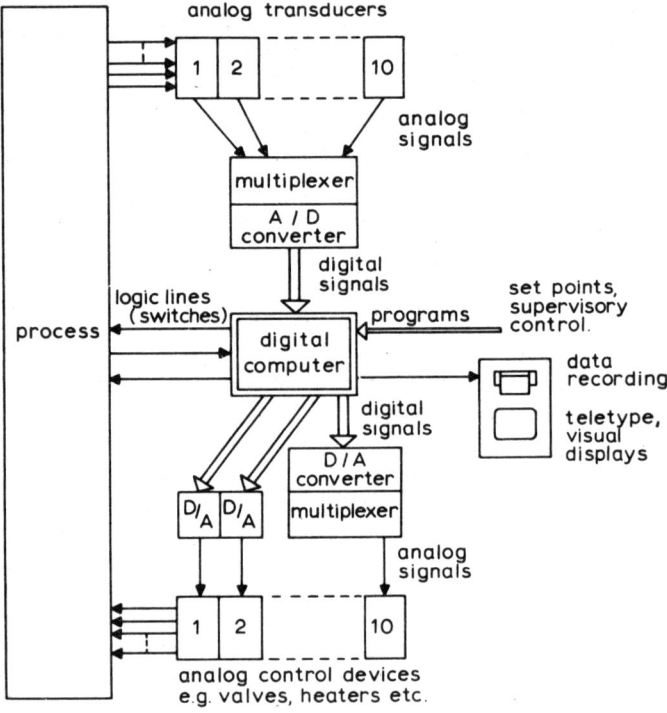

Fig. 1.2   Online digital control system

transducers produce electrical voltages proportional to the parameters to be measured. As the computer can only accept one measurement signal at a time, the transducer signals are taken in order, i.e. multiplexed, as in a data logger, before being converted in the A/D converter and read into the computer. The digital outputs to various control devices which control the process are first fed from the computer in a prescribed sequence. Each is converted to an analog signal by the digital-to-analog (D/A) converter, and then switched as analog signals to the appropriate control unit by an output multiplexer. D/A converters can be made cheaply if high resolution is not required, and so some outputs may have their individual D/A converter. In most control systems there will also be a number of direct digital lines between the computer and the process, for information which effectively requires switching actions only.

The input data to the computer is processed by the computer under the action of the programs held in store and by external commands on teletypes and similar equipment. The programs can also stipulate that relevant data, accumulated from the input signals or computed from these signals, be presented as graphical displays, typed data or on chart recorders.

An instrumentation system follows the same pattern as the control system except that there are no outputs from the computer to the process and no control devices. Generally speaking, a computer would be incorporated into an instrumentation system only where the amount of data to be collected is very large, where processing of this data is complex or time consuming and where it would probably be necessary to use an off-line computer. There is obviously considerable advantage in having a system which collects the necessary measurements and processes them and presents the data in the form required, especially if, as in a computer based system, the process can be readily changed by inputting a new program to the computer.

An alternative approach which avoids the necessity of A/D converters and, in some cases, complex multiplexers, is to use direct digital transducers, in which the parameter to be measured is converted directly into a digital signal which can be read by a computer.

It follows that if computers are to be used as on-line controllers of systems, or built into instrumentation systems, there is a growing need for digital transducers so that the input data are presented directly in a form that can be easily read by the computer. However, there is a long history of experience and expertise of analog transducers and measurement systems, and most systems will, for many years to come, have some of their measurements in analog form. This is particularly true in the process industries, where process time constants are relatively long. In avionic systems, system time constants are relatively much faster, and the control of aircraft, missiles or fire-control systems is very complex. Computers with inputs from digital transducers are used to a great extent in these applications.

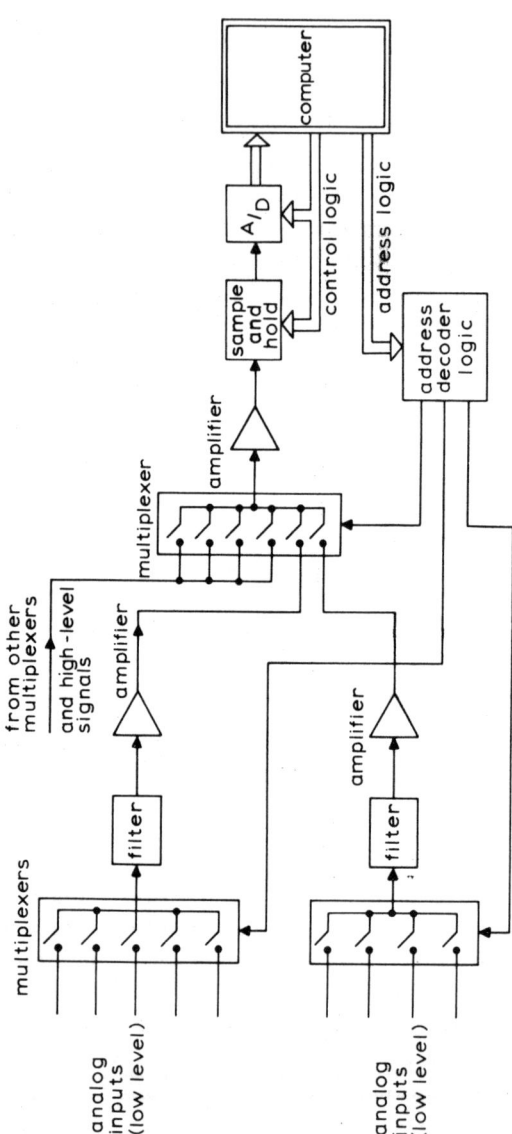

Fig. 1.3 Two-level multiplexing

Analog systems and multiplexers

In analog/multiplexer systems (Fig. 1.3), in which only one A/D converter is used, it is necessary to condition each of the analog signals to ensure that it is free of noise and is of a sufficient level so that the A/D converter can convert the signal to a digital signal of the required resolution. For example, consider an A/D converter with a nominal 10V maximum input that is converted to a 10 bit digital number. The least significant bit represents about 50 mV. Therefore, input signals of the order of millivolts, as may be obtained from thermocouples used to measure temperature, would either not be measured at all or would have only very approximate digital values. A further problem, particularly associated with low input levels, is the noise which is collected by all transmission lines and appears at the A/D converter as an unwanted voltage.

The system must therefore include amplifiers and filters and, in some cases, multiplexers at different signal levels. Such a system is shown in Fig. 1.3, the operation of which is as follows. From a program command in the central processor unit of the computer, a particular input channel is to be measured. This channel is identified by an address, which is represented by a digitally coded signal on the address lines. This is decoded to cause the appropriate switch (or switches) on the multiplexer (or multiplexers) to close, which allows the chosen analog signal to pass through a filter (to reduce the effect of noise) and an amplifier to a sample and hold circuit. At a given instant, successive signals from the computer first initiate the sample and hold circuit to sample and store, or hold the magnitude of the input signal at that instant. Any subsequent change in the input signal does not change the value stored in the sample and hold circuit. The A/D converter is then initiated to convert the sampled value into a digital equivalent which is read into the memory of the computer. The multiplexer, sample and hold circuit and A/D converter is then cleared and ready for the next sequence of commands from the computer to read any other selected analog input signal.

Input signals that are likely to vary rapidly must be sampled more often than slowly varying signals. If it is required to detect changes, it is generally necessary to sample the signal not less than twice (and preferably more than five times) the highest frequency component which it is required to be detected. Multiplexers to achieve high sampling rates are more expensive than those which sample more slowly. Multiplexers themselves often introduce noise into the system.

8 Digital systems

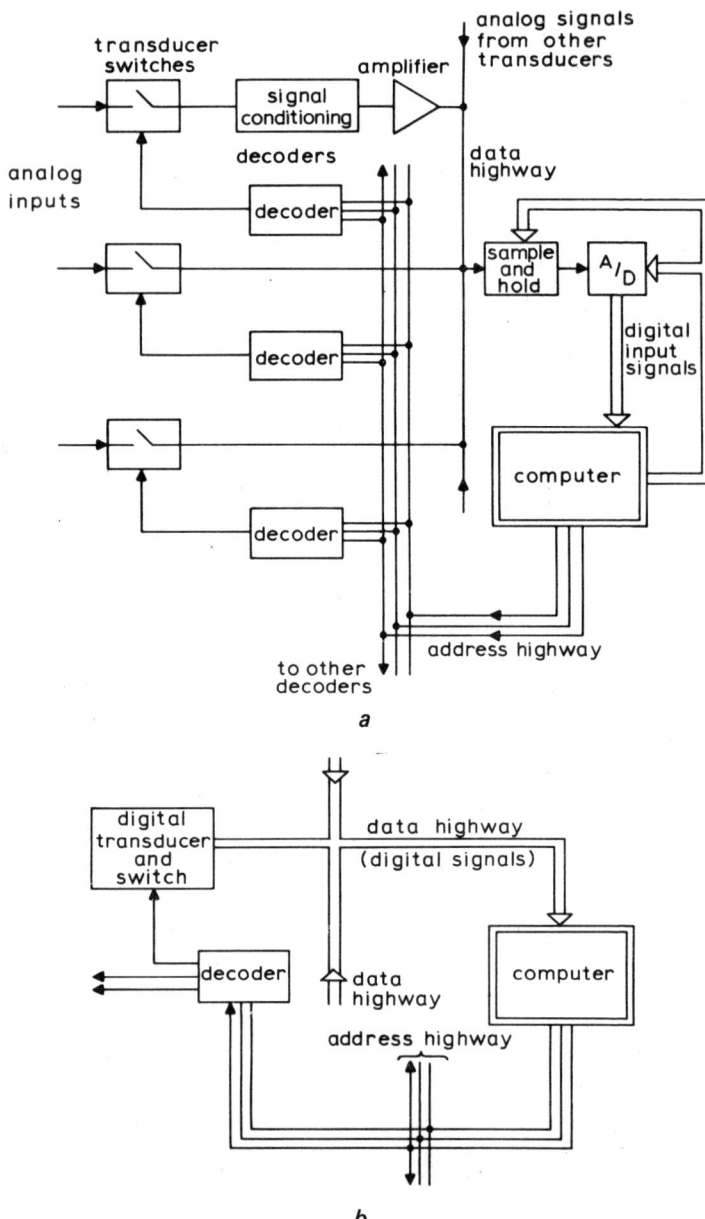

Fig. 1.4 Highway system
a Analog transducers   *b* Digital transducers

Highway systems

An alternative system to that described above makes use of a group of parallel data and address highways. Two of the many systems in operation are shown in Fig. 1.4. The first (Fig. 1.4a) uses an address highway which is a multicore cable carrying the address signals in parallel digital form to identify the transducer to be read. Each transducer carries a decoder which only switches its associated transducer to the single data highway when the correct digital code (address) is received. Each transducer signal may require conditioning and amplifying before being passed to the A/D converter, possibly through a separate sample and hold circuit.

This system may be cheaper than an analog multiplexer when the total number of transducers in the system is small. The single wire data highway can cause problems since all leads from all transducers are connected together, and interaction could increase the overall signal/noise ratio.

The second method, shown in Fig. 1.4b, uses an address highway as before, but the output of each transducer is now in digital form. A multicore cable is routed to all transducers, the actual signal on the conductors being supplied only by that transducer which has been addressed by the address highway. The transducer output must now be of the correct digital form, which can be achieved by the separate A/D conversion of the analog output at each transducer or by using direct digital transducers where possible.

Some economy on cables can be made by using a single wire address line on which serial digital data is transmitted, or by using pulse-width or other form of modulation for the address. This requires a more sophisticated decoder at each transducer than the relatively simple decoder addressed by parallel logic.

A third possible arrangement is to use analog transducers with appropriate conditioning and amplifying circuits, convert each to a digital signal in an A/D converter (under the control of a computer), and then to multiplex the digital output of each into the computer. This requires a digital multiplexer to switch simultaneously all lines of the parallel digital signal. This is often easier and cheaper to achieve, without affecting the quality of the signal, than is possible in analog multiplexers (Fig. 1.5).

The relative merits of these systems depend on the nature of the system involved, its environment, the distance of the transducers from the computer, the number and signal levels of the transducers and, of course, on the costs of the components relative to installation costs.

10 Digital systems

Drastic reductions in the price of A/D converters and associated equipment has created great interest in the system involving an A/D converter for each transducer (Fig. 1.5). However, integrated-circuit techniques allow the development of high-quality analog multiplexers

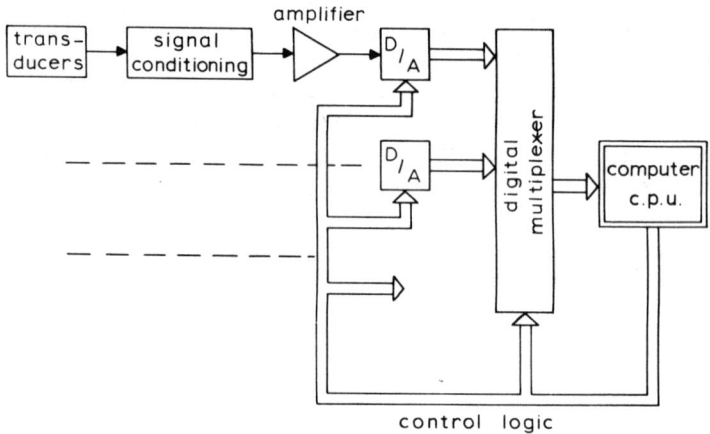

Fig. 1.5 Digital multiplexing

at costs which have also been falling. Also, the cost of wiring has increased steadily, and often rapidly, particularly where high-quality multicored cables are used. These latter points could negate the advantages of the reduced costs associated with A/D converters.

Analog and digital transducers

As mentioned previously, considerable experience has been accumulated with analog transducers, signal conditioning, A/D converters etc., and it is natural that the majority of current systems tend to use these techniques. However, there are a number of measuring techniques that are essentially digital in nature, and which when used as separate measuring instruments require some integral digital circuitry, such as frequency counters and timing circuits, to provide an indicator output. This type of transducer, if coupled to a computer, does not necessarily require the same amount of equipment, since much of the processing done by the integral circuitry could be programmed and performed by the computer.

Collins (1968) classifies the signals handled in control and instrumentation systems as follows:

(i) *Analog,* in which the parameter of the system to be measured although initially derived in an analog form by a sensor, is converted to an electrical analog, either voltage or current. Some averaging usually occurs, either by design or inherent in the methods adopted;

(ii) *Coded-digital,* in which a parallel digital signal is generated, each bit radix-weighted according to some predetermined code. These are referred to in this book as direct digital transducers;

(iii) *Digital,* in which a function, such as mean rate of a repetitive signal, is a measure of the parameter being measured. These are subsequently referred to as frequency-domain transducers.

Some analog transducers are particularly suited to conversion to digital outputs using special techniques. The most popular of these are synchros, and similar devices which produce a modulated output of a carrier frequency. For ordinary analog use, this output has to be demodulated to provide a d.c. signal whose magnitude and sign represents any displacement of the transducer's moving element. Although it is then possible to use a conventional A/D technique to produce a digital output, there are techniques by which the synchro output can be converted directly to a digital output while providing a high accuracy and resolution, and at a faster rate than is possible in the A/D converter method.

Direct digital transducers are, in fact, few and far between, since there do not seem to be any natural phenomena in which some detectable characteristic changes in discrete intervals as a result of a change of pressure, or change of temperature etc. There are many advantages in using direct digital transducers in ordinary instrumentation systems, even if computers are not used in the complete installation. These advantages are:

(a) the ease of generating, manipulating and storing digital signals, as punched tape, magnetic tape etc.;

(b) the need for high measurement accuracy and discrimination;

(c) the relative immunity of a high-level digital signal to external disturbances (noise);

(d) ergonomic advantages in simplified data presentation (e.g. digital readout avoids interpretation errors in reading scales or graphs);

(e) logistic advantage concerning maintenance and spares compared with analog or hybrid systems.

The most active development in direct digital transducers has been in shaft encoders, which are used extensively in machine tools and in aircraft systems. High resolution and accuracies can be obtained, and these devices may be mechanically coupled to provide a direct digital output of any parameter which gives rise to a measurable physical displacement. For example, a shaft encoder attached to the output shaft of a Bourdon tube gauge can be used for direct pressure measurement or temperature measurement using vapour pressure thermometers. The usual disadvantage of these systems is that the inertia of the instrument and encloder often limits the speed of response and therefore the operating frequencies.

Frequency domain transducers have a special part to play in on-line systems with only a few variables to be measured, since the computer can act as part of an A/D conversion system and use its own registers and clock for counting pulses or measuring pulse width. In designing such systems, consideration must be given to the computer time required to access and process the transducer output data.

1.4 Reference

COLLINS, G.B. (1968): 'Signal handling and computer interfaces' *in* 'Industrial techniques for on-line computers'. IEE Conf. Publ. 43, pp. 2-4

Chapter 2

Angular digital encoders

One of the most positive and direct methods of measuring shaft position is by the use of an angular digital encoder (ADE). Modern encoders offer higher resolution, greater reliability and greater accuracy than any present analog transducer of similar size. ADEs providing absolute position with a resolution better than 1 millionth of a revolution are available. This is achieved in encoders providing a 20 bit or 21 bit binary output and can represent an angular resolution of less than 1 second of arc. The high accuracies are usually achieved by using precision encoding discs and optical or photoelectric techniques.

There are, however, many encoders which are not capable of such a high degree of resolution but which nevertheless find applications in situations where the precision ADEs would prove too expensive. Such encoders may use optical techniques, but they usually have physical contact with the encoder disc or perhaps magnetic or some other form of detection.

The use of conventional analogue transducers to measure the angular position of a shaft coupled to an A/D converter generally imposes limitations on resolution and accuracy. The most common system using this method provides a voltage whose amplitude is proportional to angular displacement. To obtain a high degree of accuracy requires that the transducer be linear and that the reference voltage to the transducer be stabilised to a greater degree than the degree of overall accuracy expected. For example, an accuracy of 1 part in 100 000 requires a reference voltage of, say, 10V to be stabilised to within 0.1 mV. This is extremely difficult to produce economically. In addition, the A/D converter will introduce its own range of errors, as explained in Chapter 5. A practical limit to the overall accuracy of an A/D conversion system is about 1 part in 10 000, and to achieve this requires very sophisticated equipment.

ADEs are of two basic types: incremental encoders and absolute

14 Angular digital encoders

encoders. Incremental encoders require a counting system that adds increments of pulses generated by a rotating disc, a sensor and also some datum from which increments are added or subtracted. Absolute encoders give a complete digital readout at any angular position and do not require a datum. Fig. 2.1 shows three typical absolute shaft encoder discs.

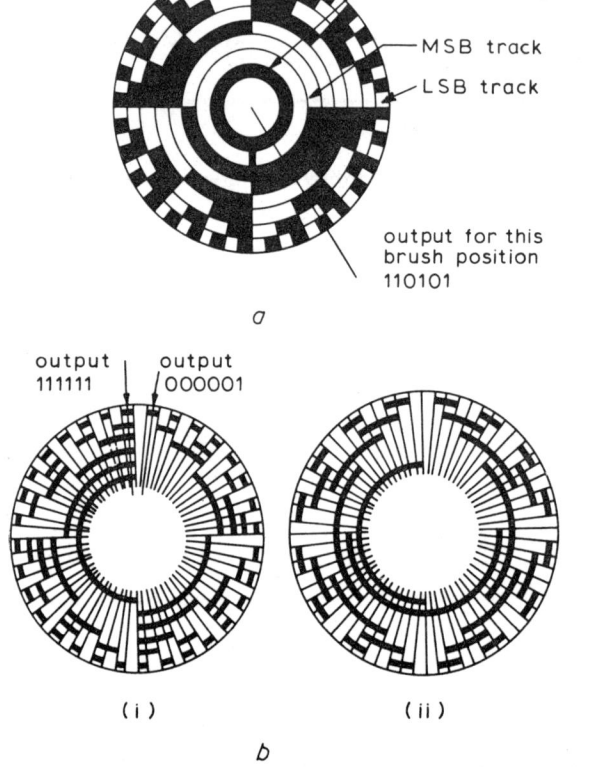

Fig. 2.1 Shaft encoder discs
    a   6-bit contact encoder disc with straight binary code
    b   6-bit optical encoder discs
          (i)   straight binary code
          (ii) cyclic binary code

2.1 Absolute encoders

Contact encoders

All absolute encoders have an encoding disc directly coupled to the angular position being measured, together with some form of sensing device. The most commonly used non-optical technique employs direct contact between the sensors (brushes) and the encoder disc. The advantage of contact encoders is that in a simple application the voltage output on each line can be adjusted to a required level by simply varying the energising supply, and no further special switching logic is necessary.

The disadvantages of brush pickups are those normally associated with brushes and rotating parts, i.e. friction between brush and encoder disc, limited life due to wear of brushes and encoder disc, particularly that due to electrical arcing as brushes move across segments, and also the inability to withstand vibration which causes lift-off of the brushes. Friction and wear can be minimised by adopting the well-established practices found in precision instrumentation. The problems of vibration can be minimised by careful design, but some conditions may still prove critical.

The encoder disc has an arrangement of metallic areas on a matrix of non-conducting areas. All the metallic areas are connected together and energised through a fixed brush, which rests on a continuous ring and is therefore in contact at all shaft positions. Other brushes are located at different radial distances from the centre of rotation of the disc, each being connected to a separate wire leading from the encoder. As the disc rotates, the brushes are connected to the common energising voltage whenever they are in contact with a metal part of the disc. Thus a lead from a brush has a voltage or logic '1' when in contact with a metallic post, and no voltage, logic '0', when on a non-conductive area.

For each position of the shaft there is a different pattern or code of energised and non-energised brushes. In a simple encoder, a single brush is used to contact one circular track on the disc. A separate track is therefore necessary for each bit of the binary output of the encoder; this is in addition to any track necessary to energise all the metallic surfaces. An alternative arrangement can be used in which more than one disc on a single shaft can be used to give a high resolution without the necessity of large discs to accommodate the required number of tracks.

Discs with up to 10 tracks each, that give a resolution of 1 in 1024, are commonly available. However, by using more than one disc and having internal gearing, overall resolutions up to 1 in 10^5 can be achieved on the electrical output. This resolution is a measure of the sensitivity of the output, i.e. a 19 bit binary number, but the total range may represent up to 4096 revolutions of the input shaft. That is, each revolution of a single disc may have a resolution of 1 in 128, but the internal mechanics of the encoder ensures that this accuracy is maintained during 4096 revolutions of the input shaft.

The fundamental accuracy of the encoder depends upon the accuracy of the disc itself, and this requirement has led to sophisticated manufacturing techniques. Many manufacturers have now developed their own techniques for the manufacture of discs.

The process generally starts with an accurate master layout of the disc, usually 5 to 10 times the actual disc size. The angles of each track that must be metallic can be computed after taking account of the various 'offsets' that are necessary to compensate for different codes, and also for the techniques used to avoid ambiguity, as discussed later. The master is reduced photographically to the final size onto a photosensitised plate, which is developed and plated to produce the metallic or conducting areas.

One of the final stages in disc manufacture is drilling the centre hole, and assembly to the shaft. It is essential that the tracks run absolutely concentric with the centre, and also at right-angles to the shaft axis. Any errors can cause erroneous counts or readings which will make the encoder less accurate than suggested by the number of tracks.

The brushes are assembled onto a block which has to be accurately positioned relative to the disc in the encoder. Each brush consists of a number of wires, each making contact with its appropriate track. Each wire of each brush has to be bent into position, since it is required to make or break contact with the metallic part of its appropriate track in a specified sequence.

Scan problems

It will be realised that an encoder having a binary output has positions where more than one of the brushes can change from contact with a metallic surface to contact with a non-conducting surface (or vice-versa) simultaneously. For example, a four-bit encoder may have a binary output 0111 which, on advancing to the next increment, will have a binary output 1000, indicating that all brushes should change simultaneously. Even with the most accurate manufacturing techniques this will not be possible. Therefore, for a minute angular displacement it is likely that one or more of the brushes will have changed as required and the others will not. There could thus be a whole range of possible outputs, all of which are ambiguous. A number of different methods have been adopted to overcome these difficulties and to ensure that all those outputs which are required to change from logic '0' to logic '1' (or vice versa) do so at the same instant and at the correct angular position of the disc. These problems do not arise if cyclic codes (such as the Gray code) is used for the disc, as only one bit changes at any instant. Thus the error is never more than the time taken for the change from a contacting to a non-contacting condition, and an ambiguous output is never generated. However, it does require additional logic to convert the Gray code to a straight binary number. Further information on these special codes is given in Appendix 1.

Alternative methods which are usually adopted incorporate two brushes on each disc, one displaced ahead (leading) and one slightly behind (lagging) the required brush position. These additional brushes are connected to leads and a logic system is used to determine the

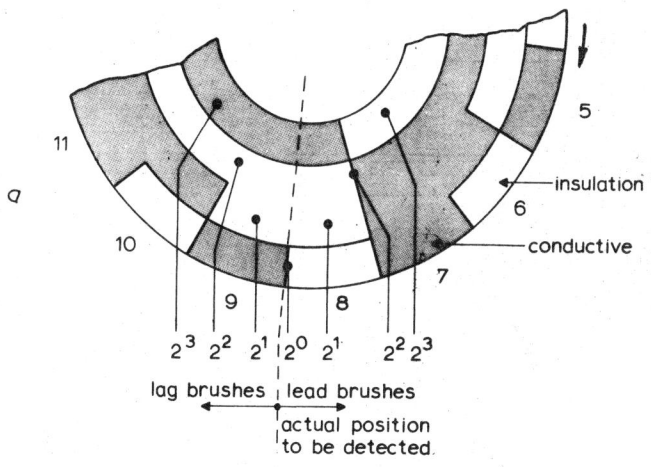

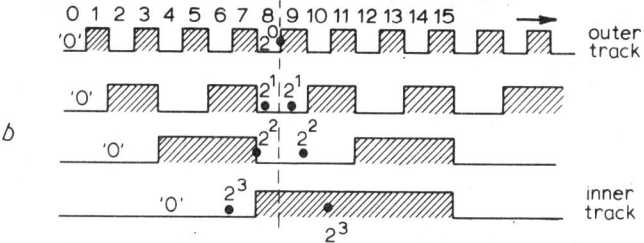

Fig. 2.2 V-scan brush positions
    a   Encoder disc
    b   Flattened layout of disc pattern

actual encoder output. A number of arrangements of these additional brushes, together with the associated logic, are in use, the most popular being known as V-scan, U-scan and M-scan. Since the LSB (least significant bit) always changes for every change in position, the LSB signal is used as a switching signal for the additional brushes. The principle involved depends on the simple fact that if a binary number is increasing, then the least significant bit changes from '0' to '1' and no other bit changes, but when the LSB change from '1' to '0' at least one other bit must also change. This is used in the V scan by locating the brushes as shown in Fig. 2.2. The associated logic necessary for the three least-significant bits is shown in Fig. 2.3.

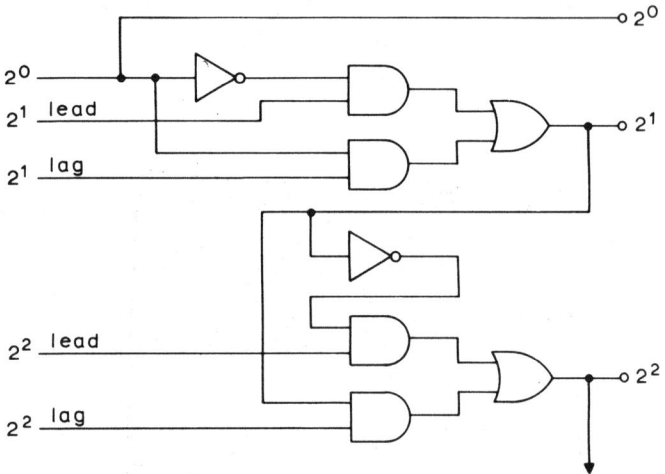

Fig. 2.3 V-scan logic

The circular disc code of a shaft encoder has been laid flat in Fig. 2.2b for convenience of explanation. Only the brush reading the LSB is retained as a single brush and this is located in the position to be measured. Other tracks each have a pair of brushes, one lagging and one leading the position to be measured. The gaps between the pairs of brushes are $1x$, $2x$, $4x$, $8x$ etc., where x is the width of one LSB increment.

Briefly, the associated logic circuitry has to identify that when the true output brush of a pair on a track reads '1', then the lagging brush of the next higher order track is to be read as the true output. If the true output from a given track is a '0', then it is the lead brush of the next higher order track which should be read as the true output. In this way the changeover of the true output from all tracks occurs only when there is a change on the LSB track. The logic circuitry will select which of the other two brushes on each track should be selected as representing the true output. This V-scan is adaptable for digital computer interrogation, since the brush selection logic can be part of the computer program.

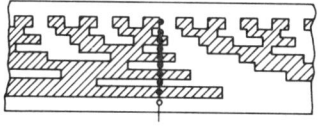

Fig. 2.4 'In-line' V-scan

In actual encoders using the V-scan, each brush often has its own track. This development has lead to a modification of the V brush arrangement which allows the brushes to remain in line, thus simplifying manufacture. With two tracks for each 'bit', the leading and lagging brushes may be placed in-line and the two tracks displayed relatively so that the two separate outputs read a leading and lagging signal relative to the true readout position represented by the LSB brush. The straightened tracks of a disc with this system are shown in Fig. 2.4. This system is sometimes termed 'V-disc' or 'in-line V-scan'.

Although this method can provide a natural binary output without

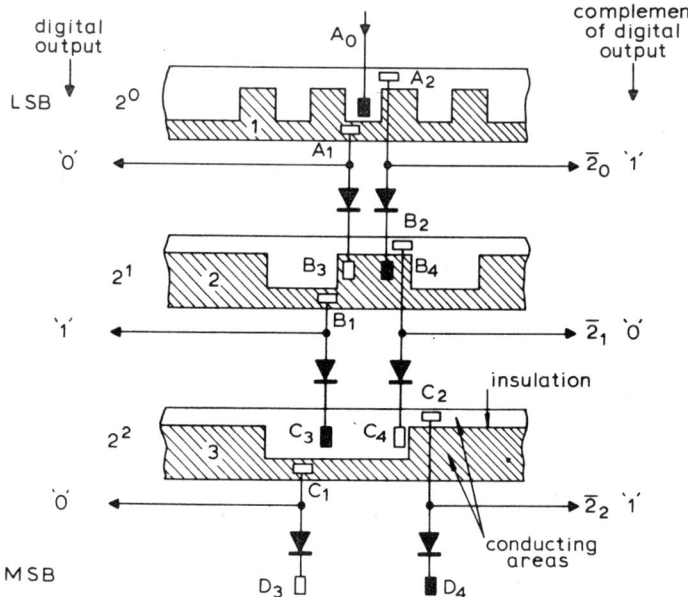

Fig. 2.5 Self-selecting V-scan

ambiguity, it does require additional external logic, and it therefore compares with the same output provided by a Gray coded disc with a single brush per track and suitable conversion logic.

A further extension of the V-scan method is a self-selecting system which requires additional brushes in the encoder and a differently patterned disc but which eliminates the need for external logic. Part of a typical disc, again straightened for convenience, is shown in Fig. 2.5. Two conducting tracks are required for each bit. These tracks are insulated from each other and from other tracks on the disc and represent the binary complement of each other. The energising voltage is fed in through brush A_0 and is transferred to subsequent tracks by a series of additional brushes as shown. Brushes A_1 transfer the energising voltage to the leading brush on the track which generates the second bit of the output, i.e. brush B_3, and brush A_2 transfers the energising voltage to the lagging brush B_4. Similarly B_1 transfers the energisation to C_3 (leading) and B_2 to C_4 (lagging). The series diodes prevent incorrect readings which could be caused by energising signals being fed backwards from C brushes to B brushes, and from B brushes to A brushes, etc.

Brush A_1 feeds to the leading brushes and A_2 feeds to the lagging brushes. Therefore, the action is the same as in the V-brush system with the necessary logic effectively being replaced by the interlocking conducting areas on each track and the energising of these tracks through interconnecting brushes. The system provides the usual logic output representing the shaft position, and also its complement, which can sometimes be of advantage in computer controlled systems. One disadvantage of this system is that the energisation for the MSB has to be derived through a series of n brushes (n = number of bits). The heavy currents create sparking and wear problems. To avoid this, drive units may be necessary between the tracks generating MSBs.

U-scan

The U-scan is effectively a combination of some of the principles of the two previous systems. The track from which the LSB is derived has a single brush while all other tracks have a pair of leading and lagging brushes. The energisation for each track is separately derived from a drive circuit. Fig. 2.6 shows the basic system.

The tracks for each bit are electrically insulated from each other as shown. A common collector track is used with each set of tracks. The LSB track acts as a switching or selector track, which is used to switch the energising supply to other collector tracks.

The Figure shows only three bits; thus the two contact areas for the second LSB (2^1), one for the leading brush and one for the lagging brush, are located against two collector tracks. There is a similar arrangement for the contact areas for the next least significant bits. This

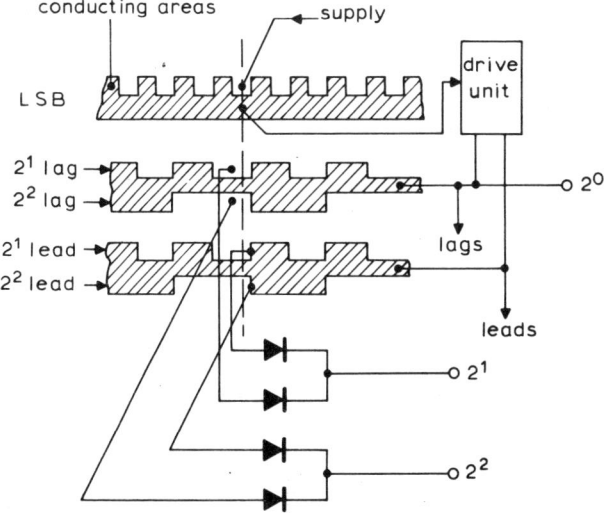

Fig. 2.6 U-scan

arrangement minimises the number of collector tracks required. Diodes between the two brushes are needed for each bit (lagging brush and leading brush) to prevent unintentional activation of the collector tracks and hence the conducting areas of other bits.

The wear normally associated with current carried by brushes, particularly as they cross from conducting to non-conducting areas (and vice versa) is minimised with the U-scan. This is due to the negligible current necessary to switch the drive circuits through the selector track. None of the other brushes carry current when passing across a conducting/non-conducting interface.

Magnetic shaft encoder

A magnetic shaft encoder has a rotating disc, coated with one of the magnetic materials used in magnetic recording equipment. The code corresponding to the conducting/non-conducting areas of contact encoders exists as a magnetic pattern prerecorded onto the disc during manufacture. An alternative technique uses a disc manufacture of suitable material which is etched to leave a raised pattern to represent the code. Generally, a magnetised area represents a logic '0' and a non-magnetised area by a logic '1'.

The pick-ups are small toroidal magnets similar to those used in digital computer magnetic memories. Coils around these magnets sense the presence or absence of a magnetic field close to the toroid. These magnets are located close to, but not in contact with, the disc. Each toroidal magnet has two coils. On one, the reading coil, the presence or absence of the magnetic field on the disc is sensed. However, an output is developed only when a second coil known as the interrogate winding is energised. Generally, the interrogation is carried out by a 200 kHz signal of constant and amplitude.

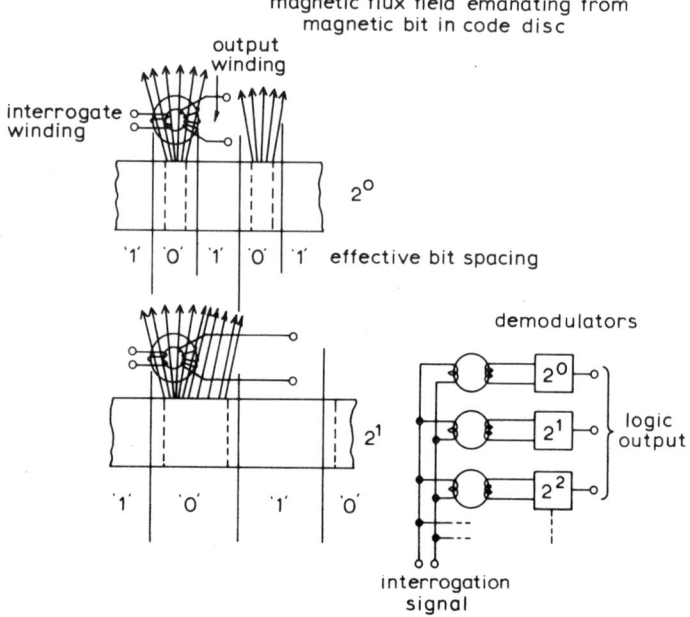

Fig. 2.7 Magnetic encoding

The output winding is therefore also a 200 kHz signal whose amplitude depends on the turns ratio and also on the magnetic field in the toroid resulting from its position over the disc. If the toroid is located over a magnetised area, then the magnetic circuit is saturated and the output voltage will be low. If the toroid is not over a magnetised area, then it behaves more like a transformer, and as the magnetic circuit is not saturated the voltage will be high.

Thus on the output coil there is a 200 kHz signal which is amplitude modulated according to the presence or absence of a magnetic area on the disc. A magnetic area causes a low output, i.e. logic '0', and a non-magnetised area causes a high voltage, i.e. logic '1'. Fig. 2.7 shows the essential features of a system of this type. Before the signals on the output windings of the coils can be utilised, it is necessary to demodulate each signal and square the resulting output to represent a logic '1' or logic '0' on the encoder output leads. Fig. 2.8 shows the type of output that is derived from, say, the LSB toroid if the disc is rotated at constant speed. The varying output levels make it necessary to include a demodulating and squareing circuit to make the signal adopt definite values representing either logic '0' or logic '1'. This is usually achieved by some form of Schmitt trigger circuit, similar to that used in optical encoders.

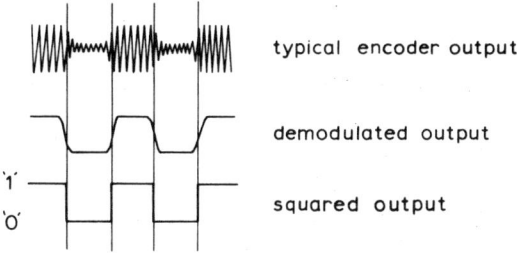

Fig. 2.8 Magnetic encoder signals

An alternative method of interrogation is similar to that used during part of a read cycle in the magnetic store of a digital computer. The interrogation winding is fed with a positive-going pulse, which generates a pulse on the output winding if the toroid is over a non-magnetised area, i.e. a logic '1'. Similarly a logic '0' (no output pulse) is produced if the toroid is over a magnetised area and the toroid is already saturated. Since the toroid is sensing the presence or absence of an external magnetic field from the disc, it is necessary to reset the magnetic field in each toroid to a given level after each positive-going interrogation pulse. This is done by a negative-going pulse immediately after the interrogation pulse.

As with contact encoders, ambiguity can exist whenever more than one bit changes simultaneously. This can be avoided by using a cyclic code or by having leading and lagging toroidal sensors, as described for contact encoders. The necessary decoding logic is the same as in the contact encoders previously mentioned.

Magnetic shaft encoders are obviously more complex than contact devices since they necessarily include expensive toroidal units and the associated interrogative and signal conditioning electronics. Experience has shown, however, that they can be very rugged and give reliable operation over a wider range of environmental conditions that can contact types. The rotational life is generally dictated by the life of the bearings, and this is considerably longer than contact encoders whose useful life is usually limited by wear and tear on the brushes.

Optical encoders

The majority of shaft encoders in use today make use of optical and photoelectric principles. In these units the disc has a pattern of transparent areas in an opaque background corresponding to the conducting and non-conducting areas, respectively, on contacting encoder discs. The complete encoder contains a light source usually coupled with an optical system, and finally an array of photo-sensing elements radially displaced. These basic elements are shown in Fig. 2.9.

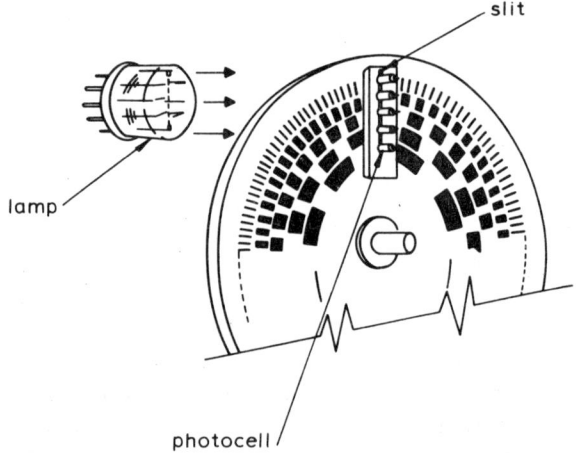

Fig. 2.9   Basic optical encoder

As in most encoders, the performance is governed largely by the quality of the disc. Most optical discs are produced photographically using a contact printing method from a master disc accurately machined to the required pattern. Much of the success of encoders is due to the specialised techniques developed for the manufacture of the master discs. The Divided Circle Machine * has produced master discs with

*used by Baldwin Electronics Inc. in the USA, and also by Vactric Control Equipment Ltd. in England

radial lines having an accuracy of 0.067 seconds of arc. This represents an accuracy of better than 1 part in 10^8. The machine can be programmed to produce a master disc to represent the more usual binary Gray codes and also to produce codes for pseudo-random, prime number, cyclic codes, sin/cos, logarithmic, and BCD outputs.

Printed discs must not only have a high degree of accuracy at the opaque/transparent transition, but must also have sharp edges. This minimises noise in the sensors during the transition between logic '0' and '1'. The sharp edges can be optically projected accurately with sharply defined shadows, which minimises edge effects. This compares favourably with magnetic encoders where edge effects create problems in the location of the toroidal pick-up units.

The discs should also have a negligible grain structure, to help the sharp changeover and keep the noise to an acceptable level.

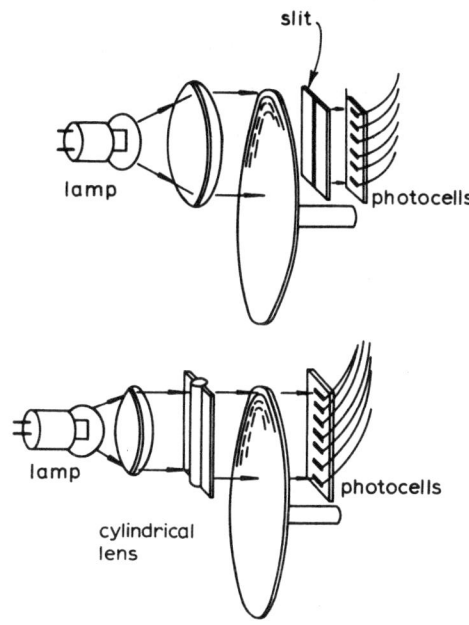

Fig. 2.10 Indexing systems

Illumination of the disc generally takes the form of a tungsten filament lamp, although solid state devices will find increasing use in the future. Fig. 2.10 illustrates the two basic optical systems at present in use. The top illustration shows a lamp and lens which is used to flood one side of the disc. The sensors are exposed through a narrow slit positioned accurately relative to the reading line.

The lower diagram shows an alternative arrangement in which the optical system produces a single line indexing beam which is projected across the reading line of the disc. Using standard lamps both these systems have been used for tracks with up to 2500 segments in one revolution. Using a special fine filament lamp, up to 5000 bits can be detected on discs of about 150 mm diameter.

Neither of the optical arrangements described above are adaptable to take advantage of the lead–lag brush systems used in contact encoders to avoid ambiguity. In practice it has been found uneconomical to manufacture complex arrangements of slits with appropriate sensors or multiple lens systems to permit a V-scan or U-scan. Therefore, many of the encoders use a cyclicly coded disc, and are required to incorporate the necessary conversion systems, if straight binary output is required.

Sensors are usually photovoltaic cells of the silicon solar type, producing about 20–40 mV in a load resistor of 10 kΩ, and amplification is required to generate logic levels of acceptable values. A separate amplifier is connected to each sensor and is designed to switch the output line to logic '0' or logic '1' at the appropriate required voltage level as the light falling on the sensor passes across the opaque/transparent interface. A nominal switching level of 10 mV + 2 mV is used for each track, produced by trimming resistors within the encoder. The output which is connected to the amplifier will therefore vary from 0 mV over the opaque areas to 20 mV over the transparent areas. Leakage through or around the opaque area can cause a small low-level voltage which is limited to 3 mV. Thus the actual output voltage swing may by 0 mV to 20 mV or 3 mV to 17 mV to maintain the correct trigger level.

Amplifiers

The function of the amplifier attached to each sensor output is to provide the desired logic level output and to switch at some predetermined trigger potential level. In some units this trigger potential may be set at some constant potential relative to the supply. This usually means that changes in the supply level will also change the trigger level. Similarly, changes in the lamp supply will effect the illumination and change the sensor output level. Using a common supply for the lamp and for generating the preset trigger level will minimise these effects, since a smaller supply will lower the preset trigger level and also the mean sensor output.

Other difficulties that arise are due to the actual output voltage of each sensor being dependant on the age of the illuminating lamp and on the temperature. Changes such as these cause a shift in the actual output voltage at the instant that triggering should take place. These latter problems can be overcome by providing a separate sensor which scans a completely clear track. The output of this monitor photocell can be adjusted by the trimming resistors and a separate buffer amplifier to the preset trigger level required for all the track amplifiers in the unit.

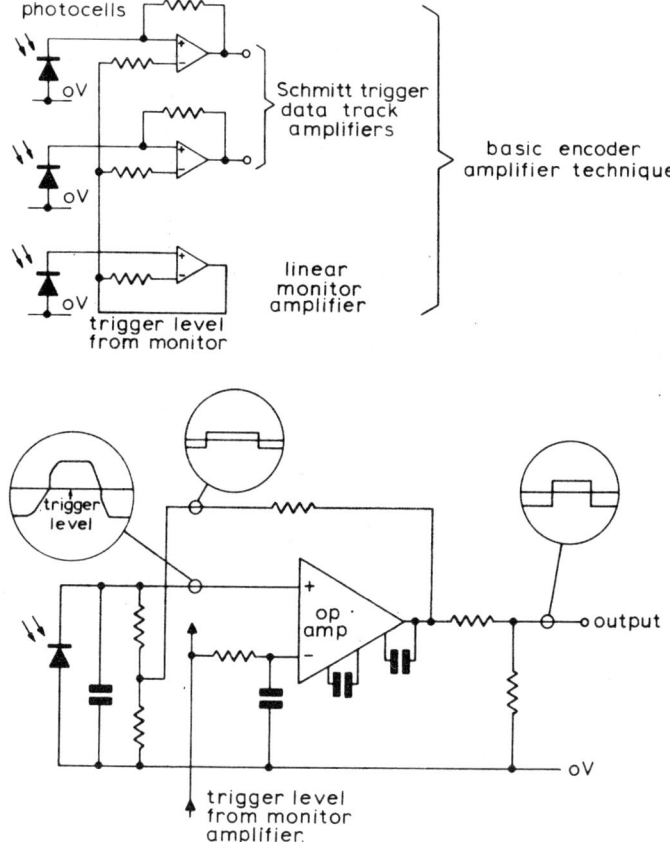

Fig. 2.11   Photocell amplifer circuit

Each sensor amplifier (Fig. 2.11) acts as a Schmitt trigger and usually consists of an integrated differential high-gain operational amplifier. One input carries the monitor or preset trigger level, and the other the sensor output. About 10% positive feedback is provided on the input from the sensor to ensure a sharp risetime in the changeover of the amplifier output, at the monitor trigger level, and at the same time keep the hysteresis effects to an acceptable level.

Hysteresis is caused by the difference in voltage level which will switch the amplifier when the pick-up is going from a dark to a clear area when the sensor output is increasing, compared with going from a clear to a dark area when the sensor output is decreasing. This is illustrated in Fig. 2.12 and will appear as a mechanical backlash in the

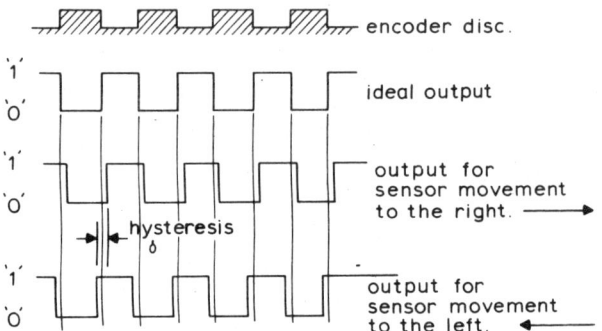

Fig. 2.12 Hysteresis in encoders

encoder output over the angle δ, as the system will indicate one of two positions depending on the direction of rotation. This can at times be an advantage, especially where mechanical vibration of a shaft can arise, in which case possible dither on the electrical output can be minimised by deliberate hysteresis.

Amplifiers can if necessary be separate from the encoder assembly, in which case precautions are necessary in transmitting the low-level signals from the sensors. More conventionally, the amplifier circuits are included in the encoder case, since the standard logic levels at the output are less susceptible to interference.

Additional facilities

Units of the type described above are available to provide complete word outputs giving the encoder's angular position at rates in excess of 400 000 words/s. The addition of electronic circuits to the basic units described above also allows developments to accommodate a variety of special applications. For example, balanced line drivers are available where the data is required to be transmitted over long distances, e.g. 1000–2000 m or more. Attempts to use a standard logic output level over this distance would make the system susceptible to noise.

In some systems it may be necessary to provide extra electronic circuitry with a store facility. In this case a 'command' pulse from a computer causes the logic circuits to store the latest value, and this is the value that can then be read when required, for example on the receipt of a 'read' pulse. This is usually a necessity in systems incorporating cyclic to natural binary conversion, to prevent a false output from the encoder which can arise during the time taken for an actual conversion. This system can also be used when a number of encoders transmit data over a common highway system, as explained in Chapter 1. Alternatively, it may be necessary to store the encoder reading just before providing a readout in serial format. Serial format can only be used where instantaneous readout is not an absolute necessity. It generally offers a cost saving in the provision of cables, and this becomes important where the information has to be transmitted over long distances. Difficulties arise at the receiving end owing to spurious noise being read as pulses, or pulses being lost owing to spurious attenuation.

Further developments are continuously being introduced to provide other special facilities in encoders. For example, some units are fitted with a special circuit to detect lamp brightness. Should this fail to reach some predetermined level, a digital signal is developed on a special output line to give warning of a possible or actual failure. It is also possible to obtain a separate logic unit which converts the actual encoder output to allow for an offset zero. This is accomplished by adding or subtracting the logic value of offset required.

The illuminating lamp is the most likely source of failure, and it is possible in some encoders to introduce a secondary optical system which can be brought into action should the first lamp fail. Although it is susceptible to failure, in practice the life of the lamp has proved to be the most predictable of all the components in an encoder system. Regular lamp replacement will usually prevent any catastrophic lamp failure, the actual operational life being determined by the operating conditions. In situations where vibration and mechanical shock are likely to occur, the life of the lamp will be reduced.

Some of the difficulties associated with filament lamps can be overcome by the use of light emitting diodes (LEDs) to illuminate the encoder disc. Encoders incorporating LEDs are currently available having resolutions up to 13 bits. Although these units have proved to be more reliable than conventional lamp units in situations involving difficult mechanical conditions, they have the disadvantage that their operating temperature range is limited.

30   Angular digital encoders

Fig. 2.13 illustrates the relative performance of filament lamps compared to LEDs, and also the performance of the usual photocell sensors, as a function of operating temperature. It is clear that a filament lamp/photocell combination has a wider operating temperature range (-40°C to 130°C) than LED/photocell combinations (-40°C to 100°C). This assumes that 50% relative output power is the limit of reasonable output.

high-vacuum filament lamp

gas-filled filament lamp

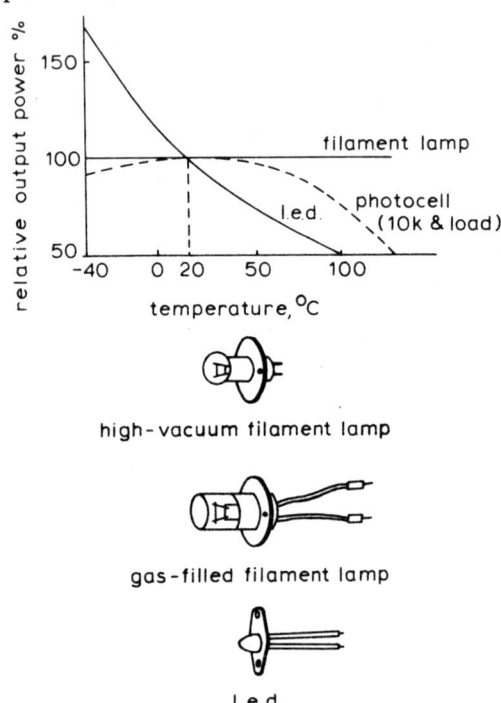

l.e.d.

Fig. 2.13 Comparison of lamp characteristics
Typical temperature performance curves

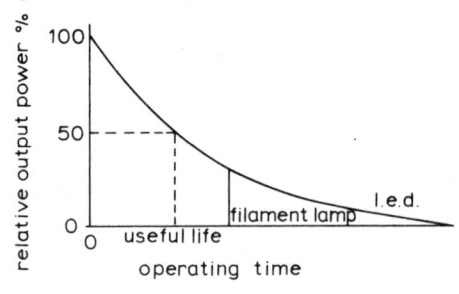

Fig. 2.14 Comparison of life characteristics
Typical life characteristics

The relative lives of filament lamps and LEDs are illustrated in Fig. 2.14. Under normal operating conditions the useful life expectancy, to 50% relative output, is approximately the same. LEDs have a longer life, but for much of this the output level is too low to be useful. Generally speaking, the operating life of filaments is extended by operating the lamp below the nominal working voltage, and this also adds to its resistance to failure by shock.

The light output efficiency of LEDs is considerably less than for most filament lamps. The output of sensors is therefore generally lower, and this necessitates more sensitive amplifiers with greater attention paid to noise and stability.

Optical resolver

The actual outputs of the sensors have a waveform which corresponds approximately to a sine wave. With a little modification in the actual shape of the track segments and the optical geometry, it is possible to produce a near perfect sine wave. The information contained in a sine wave is quite considerable, and with suitable circuits this information can be extracted. Thus, a fine track designed to produce sine wave output can be used to develop two or more subdivisions, thus increasing the resolution.

This technique is now being used to increase the resolution of ADEs in which the most significant bits (up to 14) are produced from separate tracks, and a special outer track issued to develop the additional resolution, giving 3, 5 or more least significant bits. The additional outer track has a number of special areas, proportional to the number of sectors on the finest of the other, or inner, tracks.

Sensors on the outer track are positioned to obtain a sine output and also a cosine output. In some cases two or more sensors are used in each case, and electronic averaging circuits are used to generate a true sine wave and a true cosine wave. In this way mechanical errors in the encoder can be more or less cancelled out. It must be remembered that the periodic 'time' of the sine wave output is the angular displacement of the shaft which produces a change of one bit on the finest of the digital tracks. For example, if there are 14 inner tracks, then the finest track will generate 16 384 bits with each transition ('0' to '1' and '1' to '0') representing 360/32 768 degrees or approximately 0.6 minutes of arc, and the resolver track will generate one complete cycle of output for each bit.

The outputs of the sensors on the special track (Fig. 2.15) are not dependent on the motion of the encoder disc. It is the shape of the segments and the arrangement of the optics that gives rise to the magnitude of the sensor, outputs, and this is related directly to the

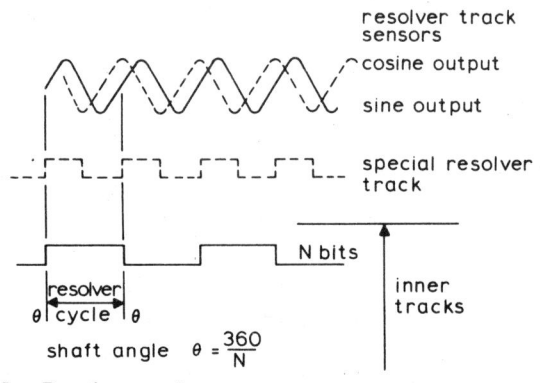

Fig. 2.15 Resolver track output

shaft position whether it is rotating or stationary. Since there is no switching action on the outputs, the lamp intensity is important and great care has to be taken to prevent lamp decay or other variations which affect the sensor output magnitude.

The interpolation network receives the sine and cosine inputs and adds them together with differing summing coefficients into a number of separate generators, producing sine waves displaced within the interval of the primary sine wave generated by the encoder sensors. Fig. 2.16 shows how the following six separate sine waves can be generated:

(a) primary sine wave
(b) 1st generated sine wave displaced by 60°
(c) 2nd generated sine wave displaced by 120°
(d) cosine wave sine wave displaced by 180°

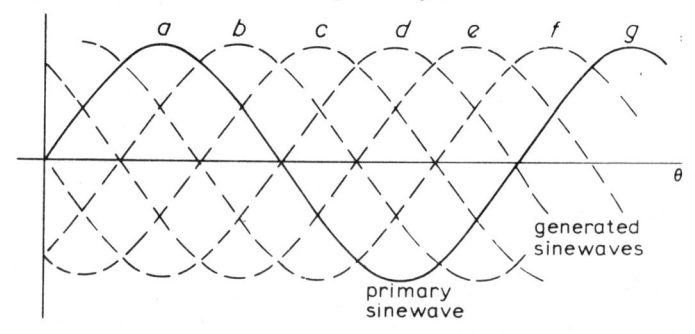

Fig. 2.16 Intermediate waveforms

(e) 3rd generated sine wave displaced by 240°
(f) 4th generated sine wave displaced by 300°
(g) primary sine wave

Each zero crossing of each wave can be used to trigger a Schmitt trigger circuit, thus producing an increase of resolution of the primary sine wave. This information can then be used to generate the additional least significant bits to represent the shaft angle. Further details of the interpolator to generate the separate sine waves and the decoding logic used to generate the addition bits is given in Chapter 4.

In practice, 16 or more intermediate sine waveforms can be produced giving 32 accurately spaced points for each of the least significant bits of the other tracks. This is equivalent to 5 additional digital bits. In an encoder with 14 inner tracks, the binary resolution can thus be increased from 1 in 2^{14} to 1 in 2^{19} and this represents a resolution of approximately 1 in 500 000 or a shaft angle of less than 2 seconds of arc. The system described above is shown diagramatically in Fig. 2.17.

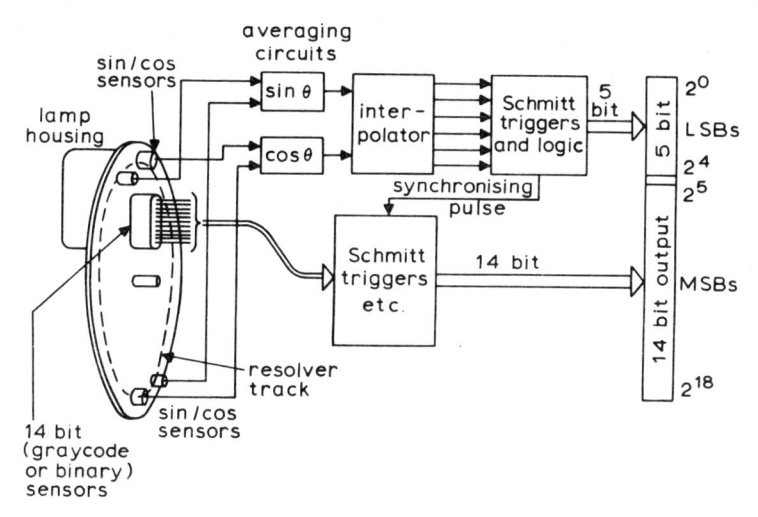

Fig. 2.17 19-bit resolver ADE

A further advantage of this resolver system is that it can be used to generate a synchronising pulse to read the inner track sensor outputs. It will be recalled that in a straight binary encoder, ambiguity can arise whenever more than one digit changes simultaneously. The techniques adopted to prevent this use the least significant changeover to decide which bits are relevant. With the resolver, then, the logic to decide when the most significant bits are read can be decided by the

34 Angular digital encoders

resolver, and this can be selected to occur when ambiguity is not likely to be present.

An alternative method to that described above is shown in Fig. 2.18. This system has the advantage that it is not as sensitive to amplitude changes caused, for example, by lamp deterioration. As in the previous system, multiple heads are used with averaging circuits, and operate on a linear track to generate sine and cosine signals; each cycle being produced over the same shaft angle as the smallest increment of the inner digital tracks.

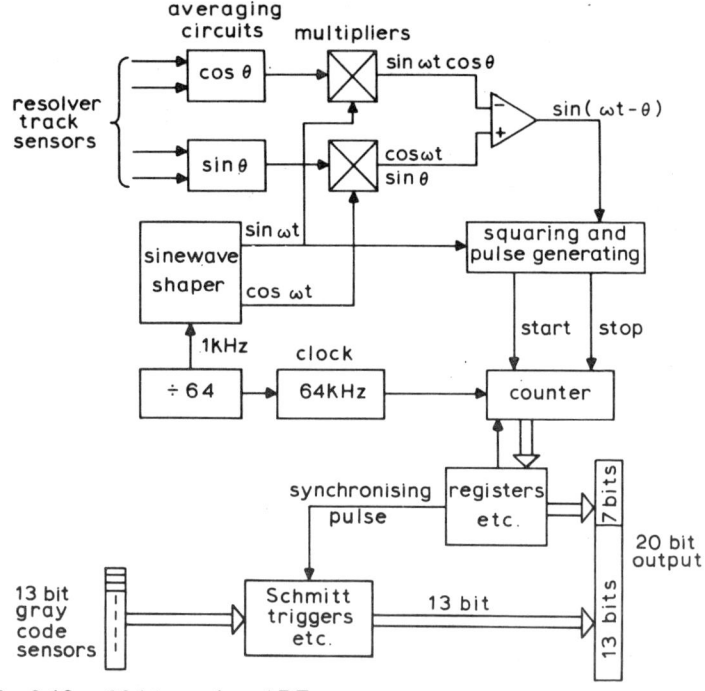

Fig. 2.18 20-bit resolver ADE

The difference in the two systems is in the manner in which the extra information is extracted. In this method, a constant frequency generator forms part of the electronics and generates sine and cosine signals at a frequency of ω rad/s, i.e. $\sin\omega t$ and $\cos\omega t$. These are two generated signals combined with the encoder sine and cosine outputs to produce two modulated signals.

$$\sin\omega t \cos\theta \quad \text{and} \quad \cos\omega t \sin\theta \quad (2.1)$$

where θ is the angle from one cycle from the sensors. These two signals are then subtracted to give a derived output signal whose phase in relation to $\sin\omega t$ depends only upon the value of θ:

$$\sin\omega t\cos\theta - \cos\omega t\sin\theta = \sin(\omega t - \theta) \qquad (2.2)$$

As θ is now a phase relationship it is completely independent of the magnitude of the sine and cosine waves generated by the sensors, and also of the amplitude of the generated signals at the frequency ω.

The sine wave generator at ω is derived from a master pulse generator operating typically at 64 kHz, which is divided down to generate 1 kHz sine waves. The derived output $\sin(\omega t - \theta)$ is used to switch the pulses from the 64 kHz generator into a six-stage counter, the number of pulses depending on the angle θ.

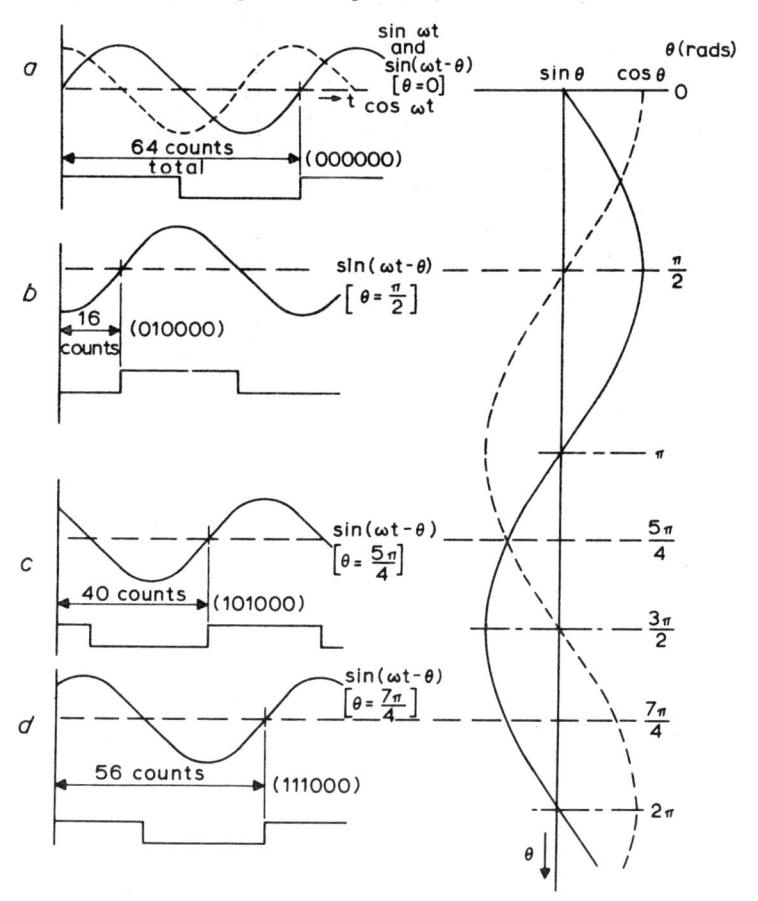

Fig. 2.19 Resolvers waveforms

Fig. 2.19 shows the relationship between the averaged $\sin\theta$ and $\cos\theta$ (θ being an angular position of the encoder disc) and $\sin\omega t$ and $\cos\omega t$, the continuous sine waves developed by the signal generator. The output of the difference network is seen as a sine wave of frequency ω but differing in phase from the $\sin\omega t$ by the angle θ. Hence when the encoder disc is in a position represented by $\theta = \pi/2$ rad, then the output is as shown by the curve (*b*) relative to $\sin\omega t$ in curve (*a*).

Squaring circuits can be used to generate square waves from $\sin\omega t$ and $\sin(\omega t - \theta)$, and pulse generators used to develop pulses only on the positive going edges of these square waves. The pulse developed from the $\sin\omega t$ is used to start the counter adding the 64 kHz pulses to the counter, and the pulse derived from the $\sin(\omega t - \theta)$ is used to stop the count. Hence, the accumulated count will represent the angle θ; 64 pulse for $\theta = 2\pi$ rad and proportionally less for smaller values of θ, as shown in Fig. 2.19.

A possible system of this type is shown in Fig. 2.18, which would function in the manner described above. In the actual system pulse interrogation is used, phased according to $\sin\omega t$ and $\cos\omega t$, and a tuned amplifier provides the waveform to control the counter store. Correlation between the most significant bit of the resolver output with the least significant bit of the Gray code output is necessary to prevent inaccuracies due to mechanical errors and the different electronic systems.

This correlation is achieved by a further track on the disc which has the same resolution as the finest track but is displaced so as to bridge the change points of the Gray code tracks. Comparison of the MSB of the resolver output and the LSB of the Gray code with the correlation track enables '1' to be added to, or subtracted from, the Gray code part of the output as required to prevent any ambiguity. The logic is similar in principle to the V-scan techniques for avoiding the ambiguity described earlier.

The frequency ω limits the maximum rotational speed for which a true readout can be obtained since the finest derived digit must not change during the 1 ms time period, required for one cycle of ω, to complete a count. Thus, the greater the resolution then the lower will be the maximum permissible operating speed. In practice, this limitation is not usually serious as high resolution is usually required only at slow rotational speeds.

2.2 Incremental shaft encoders

This section deals with the use of incremental encoders used for determining, in digital form, the instantaneous angular position of a shaft relative to some datum. Incremental techniques are also used for measuring speed etc., and these are discussed later in this chapter.

All incremental encoders are designed to generate a fixed number of pulses for each unit of angular rotation of the encoder disc. The accuracy will thus depend largely on the accuracy of the disc itself, but also on the mechanical assembly and the sensors. The actual sensing of the increments may be through contact magnetic or optical systems. Most of the techniques developed for systems used in absolute digital encoders have applications to incremental shaft encoders. For example, it is usually necessary to have some form of squaring circuit for magnetic and optical sensors in order to give a more definite indication of change of state as the disc rotates by one increment. The hysteresis associated with this type of circuit can serve to prevent false counts caused by noise in the sensors, as they pass through the transition state.

The electronic circuits necessarily associated with incremental position encoders will include counters that provide a digital output, in the required code, proportional to the number of increments generated by the encoder disc from some datum position. It is necessary, therefore, for most systems to include on the disc some indication of this datum point. Also it is necessary to generate an indication as to the direction of rotation, to identify whether any additional increments represent a clockwise or an anticlockwise rotation from the datum position. However, in subsequent discussion an optical system will be referred to, and the principles will also apply to other sensing methods.

Position logic

The upper diagrams of Fig. 2.20 show part of a typical straightened disc, the actual incremental track being the outer ring, and the single marker segment is on the inner, or datum, track. The second track contains the same number of segments as the outer track but is displaced by one half of one segment. Assuming that one 'cycle' is represented by a change of two bits (i.e. two transitions) then the second track is displaced by 90 (electrical) degrees.

The output of the outer track will lead the output of the inner track for one direction of rotation, and lag the inner track for the opposite direction. Relatively simple logic circuitry can determine the direction of rotation from these two tracks. The output pulses from the outer track are fed to a counter and can be made to count up or count

Angular digital encoders

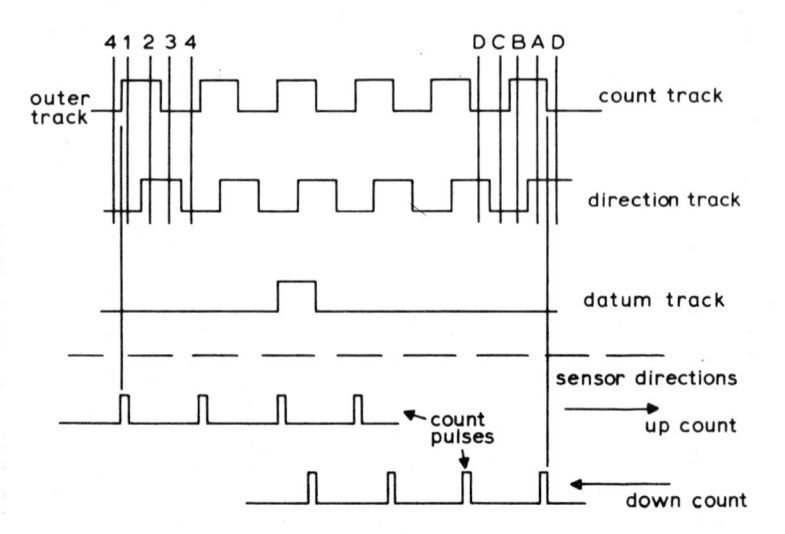

Fig. 2.20 Incremental encoder timing diagram

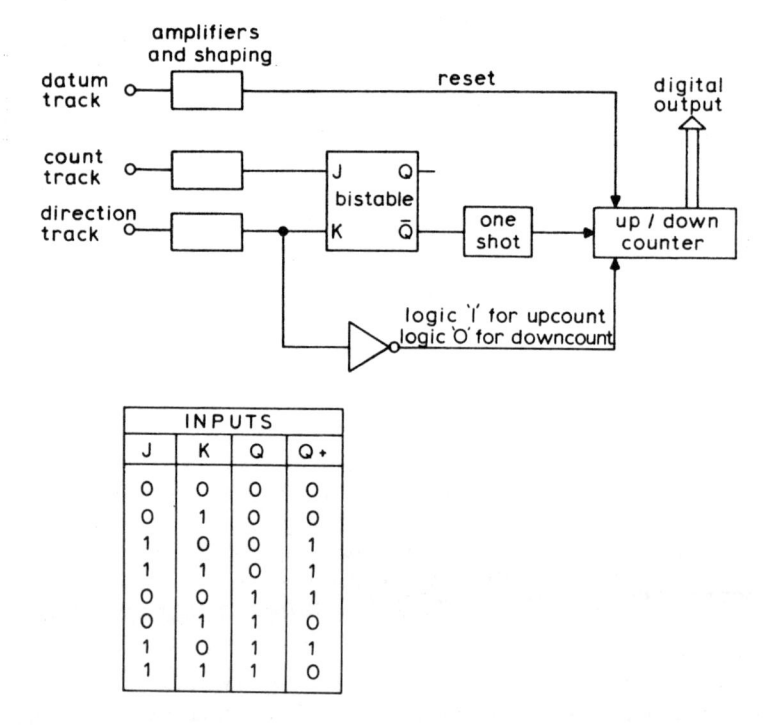

Fig. 2.21 Incremental encoder circuit

INPUTS			
J	K	Q	Q+
0	0	0	0
0	1	0	0
1	0	0	1
1	1	0	1
0	0	1	1
0	1	1	0
1	0	1	1
1	1	1	0

down according to the direction of rotation. The pulse from the datum can be used to reset the counter to zero.

Fig. 2.20 also shows the type of output that might be derived from the two outer tracks after the signals have been squared by Schmitt trigger circuits. The associated logic which could be used is shown in a simplified form in Fig. 2.21. The actual counting pulses are developed by the bistable or flip-flop output resulting from the set (J) and reset (K) inputs from the count track and the direction track, respectively.

Consider an 'up count' in which the relative motion of the sensor moves from left to right. The transition of the output from the count track from logic '0' to logic '1' is applied to the J input of the bistable and will trigger the output Q of the flip-flop to logic '1' only when the K input from the direction track is at logic '0'. This situation occurs in the 'up' count direction only. The flip-flop must be reset by a transition of the 'direction' output from logic '0' to logic '1'. Thus, the 'up' counting pulses to the counter are only initiated if there are alternating 'count' and 'direction' pulses fed to the flip-flop. This process can be verified by examining the instances labelled 1, 2, 3 and 4 with the truth table. The fact that the direction pulse is negative at the instant that the bistable changes state is used to pass an 'up count' signal (logic '1') to the counter itself through an inverter. In the 'down count' situation, the one-shot output pulse is also generated by the 'count' pulse moving from logic '0' to logic '1', but this now occurs when the 'direction' pulse is at logic '1'. The counter will at this instant receive a 'count down' command. The logic can be verified by checking the position A, B, C and D with the truth table, and to establish that no ambiguity exists.

The actual logic circuits used in industrially available counters may differ from that shown in Fig. 2.21 in order to provide a high degree of accuracy and reliability.

In practice, the one-shot pulse is usually of very short duration, since the total time of the pulse and its risetime and recovery time must be less than represented by 90 electrical degrees. The pulse width (in time) of the one-shot output governs the maximum angular velocity of the encoder.

Typically the one-shot pulses have a width of $4\mu s$ to $6\mu s$, with risetimes and recovery times of 200 ns. For an encoder with 1200 pulses per revolution this would represent a maximum speed of 5000 rev/min and 2000 rev/min for 5000 pulses per revolution.

The two incremental tracks, displaced at 90 electrical degrees, provide an opportunity to increase the resolution of the encoder by using the leading and trailing edges of both tracks to develop the 'one-shot' pulses for the counter. The use of both edges of the outer track will allow the number of pulses per revolution to be twice the number of segments. The use of both edges of the direction track as well as those of the outer tracks will increase the number of pulses to four times the number of segments. Thus a disc with 5000 segments (the maximum currently available on a 150 mm disc) can have its resolution increased to 1/20 000 of one revolution, which is approximately 1 minute of arc.

Use of synchros as incremental encoders

A method of obtaining digital output from transducers that is simpler to implement than the systems already described has been described by Hassan and Stephanos (1971). The method does not use either the voltage amplitude or phase to determine the position, and hence the relative positions of a synchro-pair are not limited to $\pm \pi/2$. More detailed information on synchros is given in Chapter 6.

The synchro is effectively used as an incremental shaft encoder since the synchro output is converted to a limited number of pulses per revolution which are counted up or down, depending on the direction of rotation. For a high degree of resolution it is necessary that the synchro rotates a sufficient number of revolutions which is achieved by proper gearing from the shaft whose position is to be determined. There are many practical situations, e.g. on position control systems, where the driving motor rotates at much higher speeds than the output shaft.

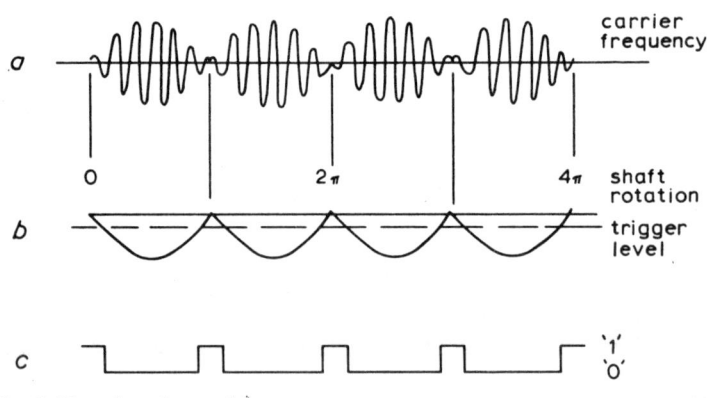

Fig. 2.22 Synchro-pulse generator
  a Synchro output
  b Rectified output
  c Schmitt trigger output

In such cases a synchro attached to the motor shaft can be used in the configuration.

A continuously rotating synchro, either the receiver of a synchro-pair or the sine output of a resolver, has an output of the form shown in Fig. 2.22. The carrier frequency is the same as the supply (or reference) frequency. The synchro output is first demodulated and rectified to obtain the wave form shown in Fig. 2.22b. This is then used with a Schmitt trigger circuit to obtain the waveform shown in 2.22c. This last signal is then used to produce pulses for the counter by using the trailing edge of each pulse to generate narrow pulses of fixed time duration independent of the speed of the shaft.

Two counting pulses will be developed for each revolution of the shaft and these are fed to the counter. It is necessary, however, to develop some additional information to indicate direction of rotation and so provide a 'count-up' or 'count-down' signal. Also, if the shaft hesitates, as occurs if the shaft stops and changes direction within half a revolution, an erroneous pulse can be developed. The logic control of the pulses to the counter must prevent this pulse being counted. The authors of the reference suggest that the direction of rotation could be determined by a tachogenerator or by some electrical network on the motor drive.

The basic logic of the circuit is shown in Fig. 2.23. The pulses from the synchro shaping circuits and their inverse are fed to the control logic which together with the direction signal generates two signals to the reversible counter to control the 'count-up' or 'count-down' functions. The pulses counted are derived from a multivibrator which generates the narrow pulses required by the counter. The generation of the narrow pulses by the trailing edge of the Schmitt trigger output and its inverse, allows the control logic to set up the reversible counter for the correct direction of count. It also ensures that the erroneous pulses

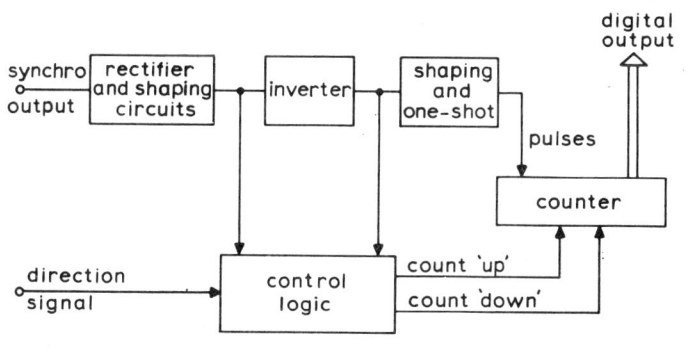

Fig. 2.23 Synchro-pulse logic

are not counted by keeping both up and down count lines at logic zero.

The erroneous pulse can occur, as mentioned, by a change in direction during the time that the Schmitt trigger is at logic '1'. The counting pulse for that Schmitt trigger output occurs on the trailing edge. If,

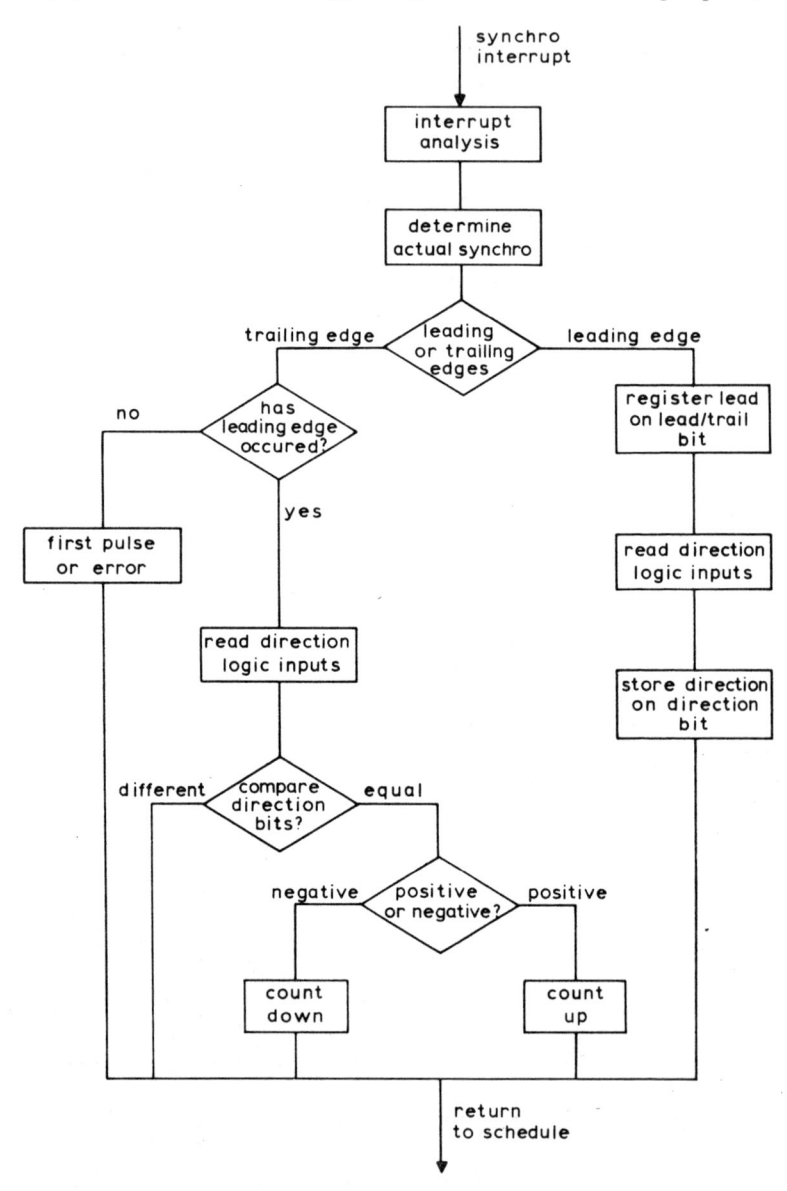

Fig. 2.24 Flowchart for synchro/computer interface

however, the synchro was moving in a forward direction and has set the Schmitt trigger to logic '1', then the counting pulse would not be generated until the trailing edge is reached. If the synchro stops and then reverses, then what was the leading edge in the forward direction now becomes a trailing edge and a reverse counting pulse will be developed. However, this must not be allowed to count-down by one as, in fact, the synchro did not completely generate the forward or count-up pulse. The timing of the pulses in the control logic takes care of this and prevents these erroneous pulses being counted.

This technique has been adapted for direct interfacing of synchros to a digital computer. In this case the functions produced by the control logic and the counter are done in the computer. Each synchro now only needs the Schmitt trigger output and a signal (logic '1' or logic '0') to indicate direction of rotation. Both the leading and trailing edges of the Schmitt trigger output are used to initiate computer action. The leading edge of each pulse is used to tell the computer to read and store the direction signal and the trailing edge is used to add or subtract from the count according to the information from the direction signal.

This latter system, although requiring a gearing if the position of a particular shaft is required to a high resolution, has the advantage of relatively simple electronics and an easy computer interface. The maximum speed of operation primarily depends on the carrier frequency, since it is necessary to develop a reasonably modulated signal for the rectifier. Also, the counting pulses, although of short duration, do occupy a finite time which must occur between the trailing edge and the subsequent leading edge of the Schmitt trigger pulses.

The flow chart for this is shown in Fig. 2.24. The computer would be operating under the control of an executive program performing, perhaps, a variety of control functions. The program is interrupted by both the leading edge and the trailing edge of the Schmitt trigger output. The computer then follows the routine outlined by the flow chart. The routine allows for a number of synchros to be connected and its first task is to identify which particular synchro has initiated the interruption. The program then proceeds as shown. The computer can perform its normal scheduled program for the relatively long intervals of time between the front and trailing edges of the input pulses from the Schmitt trigger.

2.3 Digital tachometers

Incremental encoders can be used for the measurement of velocity by either counting the pulses over a given period of time or by using the separate pulses from the encoder disc to 'gate' pulses from an oscillator into a counter.

The first method, counting encoder pulses, sometimes called 'linear function', provides only an average speed. The accuracy depends on the accuracy of the clock period and the resolution varies with speed. For example, consider an encoder disc with one hundred segments and a counting period of 6s. At a speed of 10 rev/min the resolution will be 1 in 100 as the counter will count the increments over ten revolutions. At 6000 rev/min the resolution will be 1 in 60 000. A very low speed measurement is not possible.

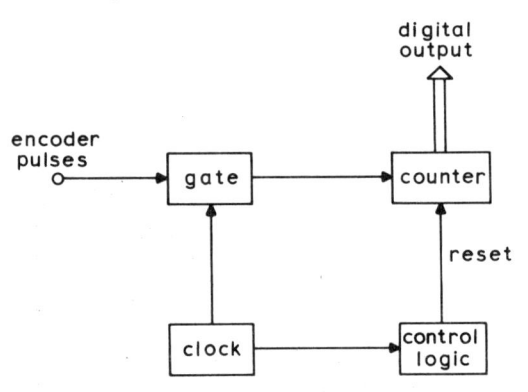

Fig. 2.25 Encoder tachometer: linear function

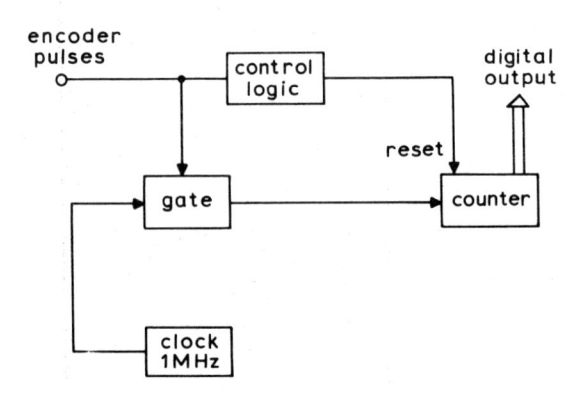

Fig. 2.26 Encoder tachometer: inverse function

The basic elements of this system are shown in Fig. 2.25. The clock provides pulses to open the gate for the prescribed period, to reset the counter before each count and simultaneously up-date the digital output. Making use of the direction detection used in measurement, the direction of rotation can also be indicated.

The second technique (inverse function) has a resolution which is greatest at very low speeds and least at the higher speeds. The technique, shown in Fig. 2.26, has a 1 MHz clock whose pulses are gated by each incremental pulse from the disc, the clock pulses going into the counter. Control logic is necessary to reset the counter and up-date the digital output at each pulse from the incremental disc.

Using the previous example, i.e. a 100 segment disc, and a 1 MHz clock, then at 10 revs/min each segment occupies 0.06 s. This gives rise to a count of 60 000 clock pulses, i.e. a resolution of 1 in 60 000. At a speed of 6000 rev/min, the resolution is reduced to 1 in 100.

The sampling time for the first system has to be decided from the speed range to be measured and the resolution required. In the second system, sampling time is dictated by the number of segments on the disc and the speed of rotation. In both systems the counter can be programmed to take account of the clock rate, the number of segments on the disc and to indicate rev/min or rad/s.

An advantage of the second method is that it does give the instantaneous speed of rotation at a given number of instances in each revolution, and it is therefore possible to detect transient changes in speed. The number of the speed indication per revolution is related to the number of segments on the disc. It is important, however, that if a single sensor only is used on the disc the segments should be equally spaced, since the clock pulses counted depend on the time the gate is opened by each segment. Unequal spacing gives rise to variations in indicated speed. Inaccuracies in the disc can be minimised by using two or more sensors equally spaced around the disc and taking each output into a logic circuit such that a pulse if generated only when any two (or three) sensors are actuated. This tends to average out disc errors.

A third technique of digital tachometry is to use an absolute encoder and a digital gating system to measure the speed. Two methods can be used. First, the encoder output could be strobed or read to give two values of encoder outputs over a given time interval. The code difference represents a given angular displacement, and this can be used to identify the speed. This would require a fairly complex digital processor or necessitate the strobing, subsequent reading and calculation to be done by a digital computer.

The second method of using an absolute encoder, shown in Fig.

2.27, is to have two digital circuits connected to the encoder output which identify two selected encoder outputs of known angular displacement. The outputs of the digital circuits are used to gate clock pulses into a counter during the time interval for the disc to output the two encoder positions.

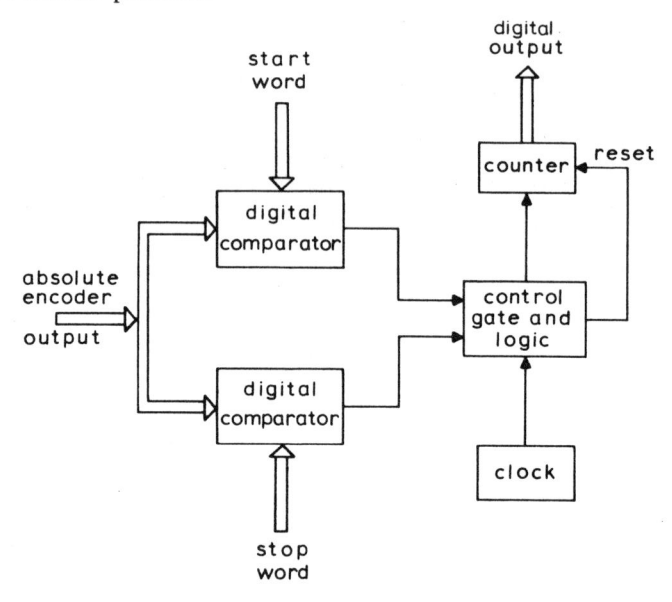

Fig. 2.27 Digital tachometer using absolute encoder

This use of an absolute encoder for measurement of angular velocity could be secondary to some other uses of the encoder. It should be noted that unless the two encoder outputs are 180° apart, the direction of rotation must be known if a true speed is to be determined.

A novel technique has been described by Szabados *et al.* (1973). It enables an accurate determination of speed over a wide speed range, including near zero, as well as giving some indication of transient angular velocity. The system is more complex than using incremental encoders and requires an additional driven element and a more complex digital system, or a digital computer, to perform the necessary calculations to determine true speed.

Fig. 2.28 shows the arrangement of the tachometer. Essentially the system measures the time taken for a given angular displacement and uses this to determine the speed of rotation. The unit contains the usual incremental optical disc driven by the shaft whose speed is to be measured. The sensor and lamp are located in a drum which is rotated at a constant known speed in the opposite direction by a synchronous electric motor.

The transducer output is a train of pulses. These are used to gate pulses from a 10 MHz clock into a counter. The counter measures the relative speed between the shaft whose speed is to be measured and the drum (or the bias speed). This known bias speed must then be subtracted to obtain the actual speed of the shaft. The bias speed can be

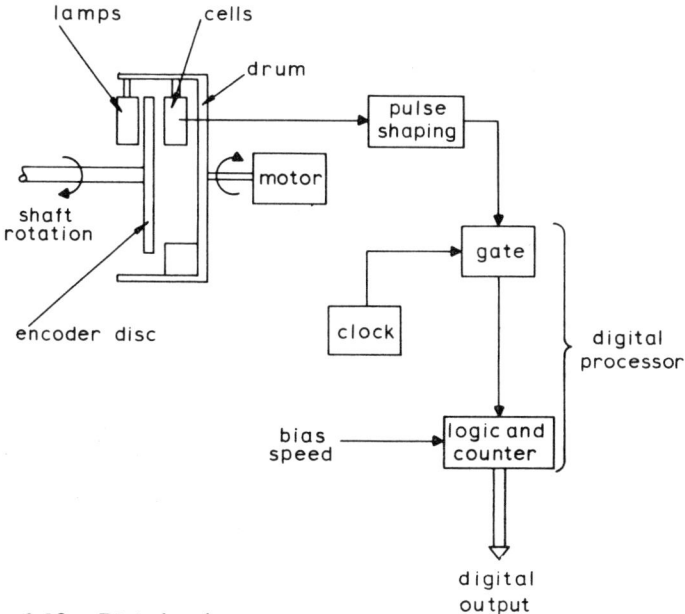

Fig. 2.28 Digital tachometer

predetermined by stopping the shaft while allowing the drum to rotate.

The accuracy of the method depends not only on the accuracy of the disc but also on the biasing speed, which should remain constant. However a model described by the authors has a resolution of 0.8 rad/s and sample rates from 1500 samples/s at zero speed up to 8000 samples/s at 7800 rev/min. The digital processor itself produces some errors, the magnitude of which depend on the complexity involved. e.g. the size of the registers, which is decided by the accuracy and resolution required. The process of subtracting the bias speed from the encoder output can delay the presentation of the true speed.

Electromagnetic pulse tachometers

In many engineering applications, particularly where only average rotational speed is required, the transducers can be formed in a cheaper fashion than is employed in optical or magnetic incremental encoders. The simple electromagnetic transducer, shown in Fig. 2.29, consists of a permanent magnet inside a solenoid, with one end of the magnet terminating in a soft-iron probe. The probe is fixed close to some projection on the rotating shaft, usually a toothed wheel as shown. As the projections pass the probe, the gap length varies and the change in flux density induces a pulse in the coil.

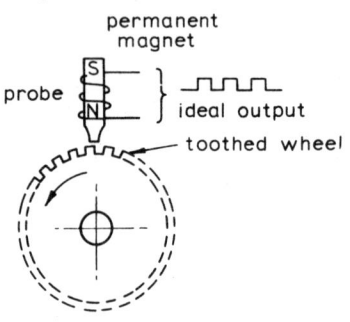

Fig. 2.29 Magnetic transducer

These pulses may be counted over a given time, or the time between pulses measured, as explained earlier for incremental encoders. Accuracy and resolution are related to the number of teeth, measurement time interval and accuracy of the teeth pitch, again as for incremental encoders.

These transducers are simple, reliable, rugged and cheap, and can be made to operate usefully over a temperature range of $-50°C$ to $+200°C$. However, owing to dependence of the output on the speed, it has been found that the limits of operating speeds are approximately from that which gives rise to pulses at 10 Hz to an upper limit of 10 kHz. Optimum performance in commercial transducers is, at the present time,

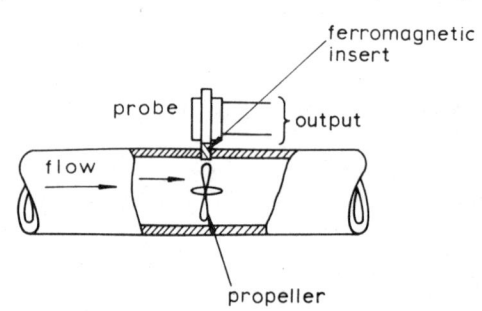

Fig. 2.30 Magnetic flowmeter

obtained from wheels with teeth pitched at about 2.5 mm and a tooth depth equal to one half of pitch.

Once the output signal has been conditioned, the digital electronics will be similar to that in incremental encoders.

A special application of this technique is used in fluid flowmeters (Fig. 2.30). A low inertia propeller is housed in a specially designed transducer and is located to carry the main stream of fluid. The solenoid and permanent magnet connect with a ferromagnetic insert that projects through the wall of the transducer as shown. As each blade of the propeller passes the insert, a pulse is generated in the solenoid as for the toothed wheel arrangement. This pulse rate is a function of the fluid velocity.

The high precision engineering associated with manufacture of the propeller assembly makes these transducers relatively expensive. The mechanical assembly requires particular attention as it is necessary that the propeller should have a minimum of drag on the fluid flow. The pulse generating system itself produces negligible back-torque on the rotating propeller.

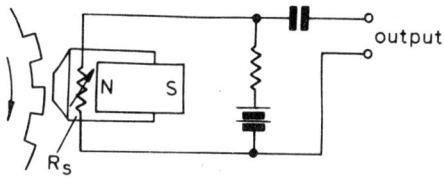

Fig. 2.31 Ferrostatic tachometer

By careful design, including the selection of materials, commercial flowmeters working on these principles have a low temperature coefficient and provide accurate and linear results over a wide range of densities. Each instrument is supplied with its own calibration curve.

The speed range of the devices using a magnetic pick-up, can be extended down to a very low speed by use of ferrostatic probes. These incorporate a resistor, R_s in Fig. 2.31, whose actual resistance depends on the magnitude of the magnetic flux surrounding it. Therefore, if this flux is changed as by the presence of a tooth of the wheel, a change of resistance takes place. These devices require an external power supply as shown. They produce an output independently of the speed of the tooth past the probe.

Capacitive tachometers

Tachometers have been developed which use the varying capacity between a probe plate and the teeth to create a pulse rate proportional to rotational speed. The gap has to be excited by an oscillator of 1–2 MHz, and so the output is effectively a pulse modulated signal at this excitation frequency. This high-frequency content can be eliminated by a capacitor across the output.

Although almost any conducting material may be used for the wheel, and practically no back torque is developed, the transducer does require an oscillator as well as amplifiers and shaping networks. The gap between the wheel tooth and probe must be very small. Any vibration between the toothed wheel and the probe causes signals to be superimposed on the transducer output.

2.4 References

HASSAN, M.A., and STEPHANOS, N.M. (1971): 'A novel application of the synchro as a digital position transducer', *IEEE Trans.*, **IECI-1B** (also *ibid.*, August 1973)

SZABADOS, B. *et al.* (1973): 'Digital measurement of angular velocity', *J. Phys. E.*, **6**

Chapter 3

Frequency dependent transducers

The use of an incremental shaft encoder to measure velocity, as given in Chapter 2, is an example of the method of digitisation that first converts the analog signal into a series of pulses at a frequency which is proportional to the parameter being measured. This technique is also used in A/D converters, and reference to this method is made in Chapter 5. Any complete system of this type will include a counter to count pulses over a given time or a timer to measure the periodic time. This chapter discusses these techniques and includes descriptions of other transducers which use the generation of a frequency as an intermediate step in measurement.

3.1 Voltage/frequency converters

The use of voltage/frequency (V/F) converters in A/D systems used in digital voltmeters and data loggers has lead to the development of a wide range of units which are available as modular units or as integrated circuits. This has led to their use in a number of transducers and instrumentation systems where an analog voltage can be developed that is proportional to a measured parameter. These units can be relatively low in cost and provide a highly accurate means of data transmission that requires the minimum of signal conditioning at the receiving end. A signal transmitted in the form of a frequency is, in effect, a serial digital signal and therefore has the advantages of digital systems but requires only two wires for data transmission. In fact, it has an advan-

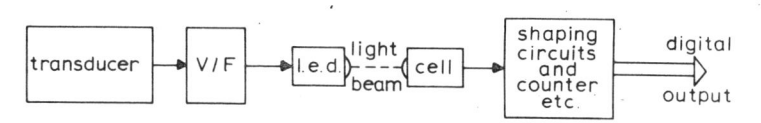

Fig. 3.1 Optically-isolated system

tage over ordinary serial digital data transmission because it is self clocking and does not need synchronisation.

A further advantage is that the technique is easily adaptable to those installations where complete physical and/or electrical isolation is required between the transducer and the actual counter or other parts of the system. This is illustrated in Fig. 3.1, in which a transducer and low-powered V/F converter are coupled to a light emitting diode (LED) whose output flashes at the frequency developed by the V/F converter. This light emission is detected by light-sensitive cells and the resulting signal shaped and counted, or computed, for the measurement. The LED and light sensitive cell are available as a single unit known as an LED optical coupler.

This form of isolation can be of value in some medical instruments. The low-powered 'front end' can be operated from a dry cell and is therefore completely isolated from the remainder of the equipment which could require a mains supply. The method can also be of value where a transducer is required to work in hermetically sealed areas, in which case the data can be transmitted across a transparent window by a separate LED with suitable screening for the light-sensitive cell.

The basic circuit of a V/F converter, which produces pulses proportional to voltage input, is shown in Fig. 3.2. In this circuit, each pulse will have the same width and height, as they represent the output of the one-shot generator. To explain the operation of the circuit let $R = 1M\Omega$ and $C = 1\mu F$. Assuming an input voltage V_i of -1 V, the capacitor will be charged through R at a rate of -1 V/s. If the comparator is set to trigger when the input is $-1mV$, then it takes only 1ms for the charge on C to reach this value. The comparator output will trigger the one-shot generator to produce a single pulse, in this case of about $1\mu s$. The 5 V one-shot output is used to initiate a circuit to allow the capacitor to discharge to earth. The charging process then recommences and the whole cycle is repeated. For the values given, the -1 V input gives rise to $1\mu s$ pulses at a rate of 1 kHz.

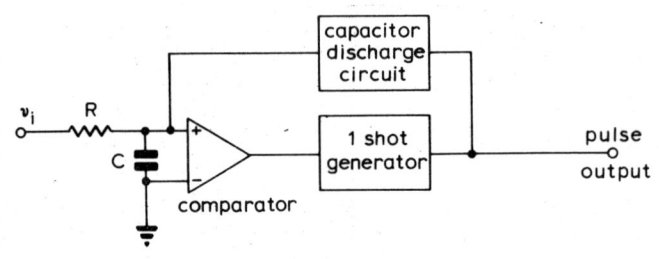

Fig. 3.2 Voltage/frequency converter

For this circuit, and an input voltage range of 0 to −10 V, the 1μs pulses are produced at rates varying from zero to 10 kHz.

The simple circuit requires considerable modification to produce a linear output with minimum changes in pulse rate, due to temperature changes. In particular, the capacitor discharge circuit calls for considerable circuitry and, therefore, a number of alternative techniques have been adopted.

A typical commercial V/F converter * is available as a module measuring 50mm x 50mm x 10mm. The unit normally produces a logic '1' output and the pulses appear as logic '0', each of 75 μs duration. The frequency range is up to 10 kHz, for a voltage input up to −10 V, with a linearity better than 0·005%. The overall accuracy and step response is such that the output can be converted to a 13 bit digital output either by counting or by measuring the time between two successive pulses. The input is arranged to accept voltage or current and allows a wide range of zero and range adjustments.

The main disadvantage of V/F converters, when used as part of a digital transducer system, is the time needed to take a reading. If a count of pulses is to be made, the accuracy can only be maintained if the period is sufficiently long to produce the required resolution. For example, for a 13 bit digital number, over 4000 pulses are required. For low-level analog inputs, which give rise to a low pulse rate, the counting period could be too long. The time can, however, be kept within reasonable limits by biasing the input signal. This can be done in the converter described above by arranging the analog input to have a range of 0 V to −5 V, and biasing the input to the unit such that its range, is, in effect, −5 V to −10 V. In this case the output frequency will range from 5 kHz to 10 kHz, and the resolution of 1 in 4000 is achieved in less than one second.

By measuring the time between pulses this problem is not so acute, but some of the advantages of the V/F technique are lost, i.e. the high noise rejection that the counting method introduces. High noise rejection is caused by the integrating action over the counting period, which produces a measure of averaging of the input signal during this period, and any common mode noise will normally have little overall effect. If the sampling period for a count is synchronised with the mains supply, the noise produced at this frequency, and its harmonics, is almost averaged out. However, noise and ripple on the analog input may alter the timing of the individual pulses, and measurement of time between pulses will now include the noise element.

*from Ampleon Electronics Ltd.

Measurement techniques

In either pulse counting or time measurements the accuracy will depend on a clock built into the measurement system. In most systems a high clock rate is chosen together with a divider network to produce other pulse rates to match the transducer or V/F converter output.

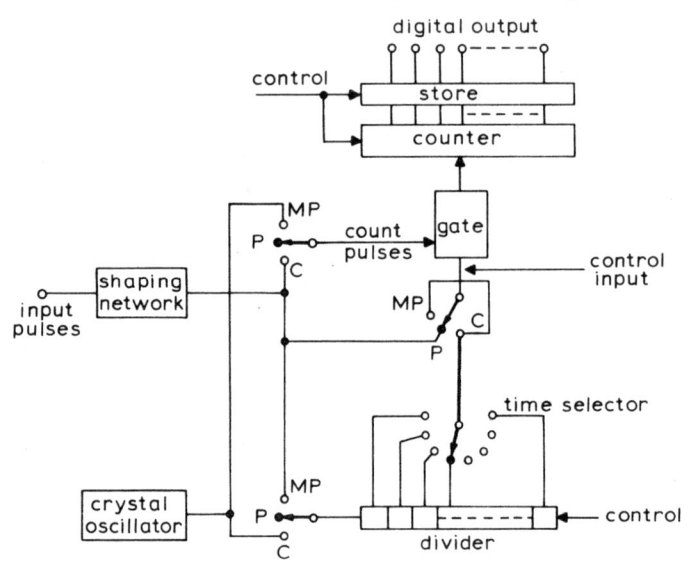

Fig. 3.3 Basic timer/counter

The basic counter/timer is shown in Fig. 3.3. A function switch enables the unit to be used as a counter (C) or to measure period (P) or multiperiod (MP). To ensure positive switching of the gate, an input shaping network is often included, as shown. In the C position, the crystal clock controls the gate to allow input pulses to accummulate in the counter for the time period set on the 'time selector'. At the end of this period some control logic is necessary to transfer the counter output to the output store, reset the counter and initiate another count to update the output.

In the period position (P), it is the clock pulses that will be counted into the counter for the period of two successive input pulses. The time selector is inoperative in this mode as the fast rate of clock pulses is needed to give the greatest resolution.

The resolution of the system as a counter can be selected by appropriate setting of the time selector. The longer time given for the counter, the greater the resolution, although the final count always represents the average rate of input pulses over the time selected. In the period position, the resolution of the periodic time measurement depends on the clock rate, and, if the periodic time being measured is short relative to the clock rate, the resolution will be low. To improve this, the unit has a multiperiod (MP) position in which the clock pulses are counted into the counter at their fastest rate but the counting continues for the number of input cycles selected by the 'time selector' switch.

For example, if the clock-rate is at 10 MHz and the input pulses occur at 100 kHz, then a resolution of 1 in 100 is the best that can be achieved in the period position. However, in the multiperiod position with the time selector switch set to open the gate for, say, 100 input pulses, a resolution of 1 in 10 000 can be achieved.

3.2 Transducer oscillators

The common feature of the devices described in this section is that a train of pulses is produced which does not depend on relative motion between parts of the transducer. The simplest form consists of an electronic oscillator in which some part of the circuit is modified by the physical property or displacement being measured, and which changes the frequency generated by the oscillator. Typical examples are changes in resistance, capacitance or inductance. All the conventional forms of analog transducer, which depend on changes of this type, can be adapted to give frequency dependent output.

With microminiaturisation in electronic circuits, it is now possible to produce very small oscillators which may be built into the conventional transducer, the output now being a variable frequency instead of a normal d.c. analog signal. Such transducers are available to measure pressure and small linear displacements and usually incorporate resistance changes in the form of strain gauges or potentiometers.

Thermistor temperature-to-frequency converter

A method described by Lövborg (1965) and others, involves the conversion of temperature to a frequency signal by means of a thermistor which is part of a frequency-determining network of an RC oscillator. The instrument operates at a low temperature, i.e. from 10 to 40°C, and is linear to within ± 0·1°C. With suitable thermistors, the range

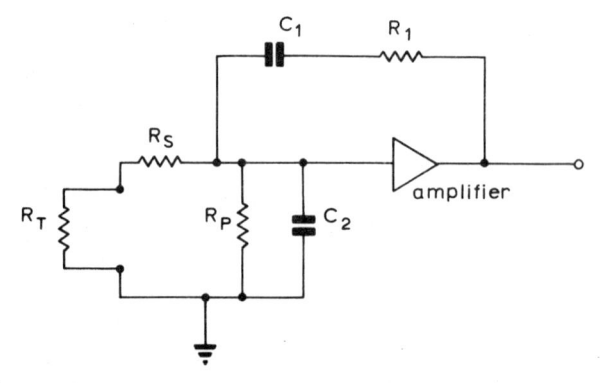

Fig. 3.4 Thermistor oscillator circuit

can be extended up to 200°C and maintain the same accuracy.

The basic circuit diagram is shown in Fig. 3.4, in which R_T is the resistance of the thermistor. This resistance is given by:

$$R_T = R_O \exp\{\beta(T-T_O)\} \quad (3.1)$$

where R_O = resistance of thermistor at temperature $T_O\,°K$

R_T = resistance of thermistor at temperature $T°K$

β = constant

The frequency of the *CR* oscillator is given by:

$$f = \frac{1}{2\pi}\left[\frac{R_P + R_S + R_T}{R_1 R_P (R_S + R_T) C_1 C_2}\right]^{½} \quad \text{hertz} \quad (3.2)$$

The resistors R_S and R_P are included to improve the linearity of the transducer. Further, the current through the thermistor must be kept to a minimum to minimise the Joule heating. The model reported in the reference (Lövborg: 1965), had a frequency range of 350 to 600 Hz as the temperature varied from 0 to 45°C. One important characteristic that is not reported is the dynamic characteristics of the transducer, no indication of the possible time constant of the system is given. Owing to the thermal capacity of the thermistor, the time constant could be longer than is normally associated with the conventional thermocouples used in most plants, although comparable to conventional resistance-type thermometers which can be used in similar oscillator-type circuits.

Quartz temperature/frequency converter

Quartz crystals have been used for some time as resonators for oscillators where stability is important. One of the difficulties is that the resonant frequency changes with temperature, the magnitude of the frequency f_T at any particular temperature T is given by

$$f_T = f_0(1 + aT + bT^2 + cT^3 + ..) \qquad (3.3)$$

where f_0 = fundamental frequency at $T=0°C$

This applies to crystals that are cut in the most suitable directions. It also accounts for the necessity of keeping these crystals in temperature-controlled ovens when used in stable oscillators. There does exist one particular cut of the crystal in which the T^2 and T^3 terms are negligible, i.e. b and c in eqn.3.3 approach zero. Over a temperature range of 0 - 200°C, this cut of crystal has a linear temperature coefficient within a few millidegrees. This coefficient is found to be 35·4 parts per million for 1°C. The basic fundamental frequency f_0 depends on the precise thickness, and hence it is possible to make a crystal which has any required temperature coefficient in terms of cycles per second per °C.

In a commercial thermometer based on this principle *, the coefficient selected was 1000 Hz per °C; hence:

$$\text{fundamental frequency} = \frac{1000}{35\cdot4} \times 10^6 \text{ Hz}$$

$$\cong 28 \text{ MHz}$$

The crystal forming the thermometer is approximately 6·5mm diameter and sealed in an inert atmosphere in a small container. The probe is rugged and can be manufactured to operate in very stringent conditions. The difficulty is that the electronic circuitry required is quite considerable as it must keep the crystal oscillating and detect changes in frequency of the order of 1 part in 280 000 if an accuracy of ±0·1°C

*Applications Engineering Group of Hewlett-Packard Ltd (1965)

is to be obtained. Further complications arise in this system as, at frequencies of 28 MHz, the type and length of cables becomes important, and also the counter circuits must be capable of handling pulses at these high frequencies.

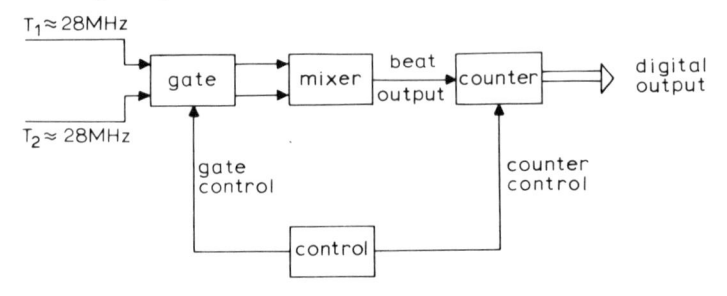

Fig. 3.5 Quartz temperature/frequency converter

A block diagram of the necessary electronics is shown in Fig. 3.5 for the case where two sensors are used. One is used as a reference and kept at a known constant temperature T_2. An alternative arrangement is to use a separate crystal-controlled oscillator (with temperature control) to act as a reference. The output of sensor T_1, subjected to the temperature to be measured, is mixed with the reference frequency and the difference or 'beat' represents the difference in temperature between T_1 and T_2 at a rate of 1000 Hz/°C. Conventional pulse-squaring and counting techniques can now be used. Certain calibration problems arise that can be allowed for in the design and these are detailed in the reference.

The operating temperature range is $-40°C$ to $+230°C$ which limits the application. The units have small thermal inertia and as such can be designed to have a very small time constant. An accuracy of $0·0001°C$ is claimed but this must depend on the pulse counting time available.

Quartz pressure/frequency converter

The same type of crystals have been adapted to serve as a pressure transducer producing a basic variation of frequency of 2 Hz for a pressure change of 1 atmosphere. In this case the crystals are cut to minimise the necessity for tight temperature control. The frequency will then depend primarily on the pressure applied across the faces of the crystal.

Hammond and Benjaminson (1968) describe this transducer in greater detail. The system works on a basic frequency of 5 MHz plus

2 Hz for each atmosphere of added pressure. A mixing circuit and counter are required in a manner similar to the temperature transducer. An added complication is that frequency multiplying is used to increase sensitivity. Without this the resolution would be limited, but, as with most other devices requiring counting, the final sensitivity depends on the time allowed for counting. For this particular pressure transducer it is claimed that for an overall range of 0 to 700 atmospheres, the resolution is 1 in 10^6 for a one second counting period and 1 in 10^7 for a ten second counting period.

The electronics involved in these devices and the cabling requirements make this and similar devices rather expensive relative to the use of strain-gauge pressure transducers and V/F converters.

Vibrating string and vibrating beam transducers

A number of transducers have been built that depend on the mechanical vibration properties of stretched strings, beams and diaphragms. The simplest of these is the stretched string that has been used as a strain gauge for load measurement for many years. The principle is illustrated in Fig. 3.6. The metal string is stretched to some predetermined load between two supports, one support being fixed relative

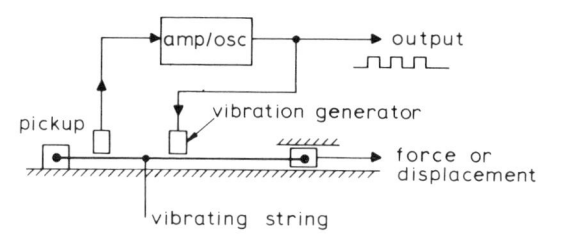

Fig. 3.6 Vibrating string transducer

to the other, to which the load or displacement to be measured is applied. Adjacent to the wire is an electromagnetic pick-up which feeds the oscillator-amplifier that provides the output and also drives the electromagnetic vibration generator. The initial vibration is usually achieved by an electromechanical device that plucks wire when a pulse is applied. The string will then continue to vibrate at its natural frequency, which depends on its length and the load or force applied. The pick-up and the amplifier-oscillator loop ensures that this vibration is maintained at constant amplitude. Any change in the load on the string will change the tension and hence its natural frequency. The amplifier output frequency is therefore a direct measure of the force applied.

Mechanical arrangements allow the device to be used over a wide range of loads but only a limited range of displacement can be accepted. In practice the units are used primarily as strain gauges on structures, and, as such, they are extremely sensitive. They are particularly useful in buildings where they are cast into concrete structures and measurements are taken periodically to assess the strain developed in the structural members. Special precautions are necessary because of the sensitivity of the natural frequency due to thermal contractions or elongations of the wire. Great care is required in construction to minimise these effects.

A development of this method has been explained by Voutsas (1963) in which a small twisted beam is used as the vibrating member, the transducer input being angular displacement which alters the angle of twist and hence the natural frequency. The transducer is small in size, fitting in a 40mm cube and has a useful operating input extending over 45° of arc. For the particular model constructed the change in frequency of the beam is found to be given by:

$$\Delta f = 24 \cdot 4984\, (1 - 0 \cdot 000059\, \Delta\theta^2)\, \Delta\theta \text{ Hz} \tag{3.4}$$

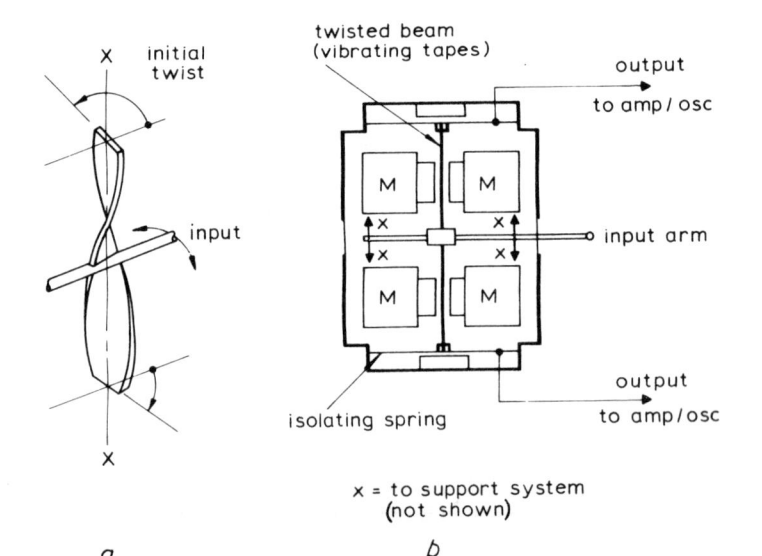

Fig. 3.7. Twisted-beam transducer
    a   Twisted beam (tape)
    b   Transducer contruction

where Δθ = change in angle of twist

0·000059 $\Delta\theta^2$ is a nonlinear term which represents a negligible error for most practical purposes. The 'beam' is in fact two thin metal tapes joined by a coupling arm (Fig. 3.7) through which the input is connected. The other ends of each tape are supported in linear isolating springs which maintain constant longitudinal tension on the tapes. Not shown is a complicated assembly necessary to support the input arm and its coupling to the tapes in order to maintain stability of the rotating assembly with respect to the housing. Rotation of the input arm causes one tape to increase its twist and the other to decrease.

The tapes themselves form part of the oscillator-amplifier circuit which is necessary to sustain transverse oscillation. The tapes serve as the transducer of the oscillator-amplifier, and their motion in the field of the permanent magnets induces a voltage which is fed back to the oscillator-amplifier. The tapes, therefore, vibrate at their resonant frequency which can be measured in the usual way to give a direct

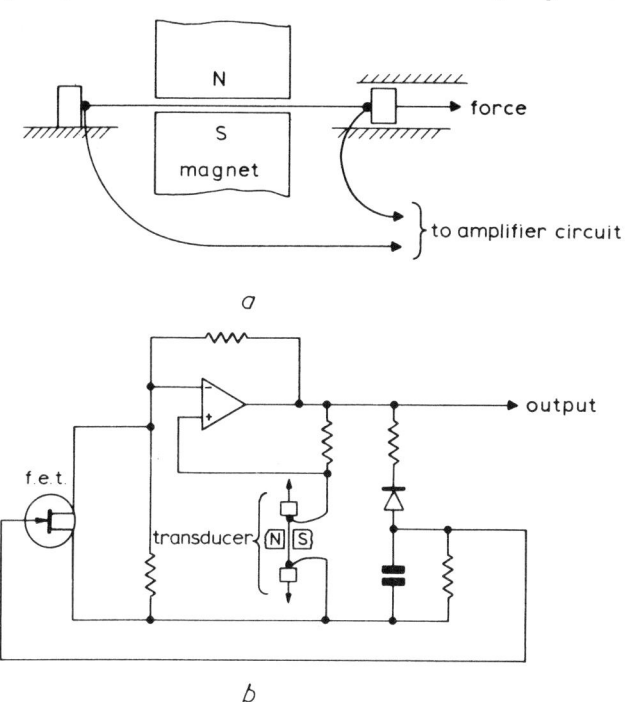

Fig. 3.8 Vibrating-wire force transducer
    a   Mechanical arrangement
    b   Circuit diagram

digital indication of the angular displacement of the input arm.

A high degree of stability is claimed and accuracies of the order of 0·0002% have been achieved over the operating displacement of 45° of arc.

A force transducer working on a similar principle has been described by Wyman (1973). This consists of a vibrating titanium wire (0·063mm diameter and 32mm long) located between the poles of a permanent magnet, as shown in Fig. 3.8a. The transducer input is the tensile force applied to the wire through insulated anchorages. As in the twisted beam transducer, the wire forms part of the electrical detecting network which maintains the wire vibrating at the frequency of its transverse resonance. The electrical impedance of such a wire vibrating in a magnetic field peaks sharply when vibrating at this frequency.

The oscillator circuit, Fig. 3.8b consists of an amplifier with broadband negative feedback and narrow-band positive feedback. The centre frequency of the narrow band is set at the mean frequency of vibration over the operating range required. The output amplitude is regulated by overall control of the loop gain by the field effect transistor. This circuit will maintain the output at a frequency equal to the natural frequency f of the tensioned wire. This frequency is given by:

$$f = \frac{1}{2L}\sqrt{\frac{T}{m}} \quad \text{hertz} \tag{3.5}$$

where L = length of wire

T = tension force

m = mass per unit length

The impedance Z of the wire at this frequency is given by:

$$Z = Q\frac{\pi\ B^2\ L^2}{12\sqrt{mT}} \tag{3.6}$$

where Q = 'quality' factor

B = transverse flux density provided by permanent magnet

The transducer described in the reference (Wyman: 1973) had a centre frequency of about 3700 Hz for a tension load of 0.81N. The temperature effects were good, as were the linearity and repeatability. Response time, as might be expected, is very rapid except that oscillating applied forces could set up unwanted longitudinal vibrations in the wire and give rise to an unreliable output frequency. As (from eqn. 3.5) this output frequency varies as the square root of the tension, then some further electronic processing may be required before a straightforward counting system can be used for a linear calibration of force.

With both string and beam systems the transducers are susceptible to 'noise' in a mechanical sense. External vibration could cause other modes to be excited which, unless specifically suppressed either by mechanical or electrical damping, would give rise to considerable errors in the output. Similarly, the dynamic performance does not seem to have been investigated and it seems likely that an ideal step response, i.e. a requirement for an instantaneous change in frequency, for a 'step-input' could not be achieved. The time constant of the transducer will depend primarily on mechanical considerations and also on the ability of the system to settle to a new frequency. High natural frequencies are necessary to ensure a fast response if a short counting time base is to be used.

Vibrating diaphragm pressure transducer

Vibrating diaphragms can also be used to measure various parameters, particularly pressure. A transducer of this type has been described by Belyaev *et al.* (1965), in which the same basic principle is used as in other vibrating transducers.

The body of the transducer (Fig. 3.9) contains a cavity, the top of which acts as a diaphragm and deforms under the pressure in the cavity.

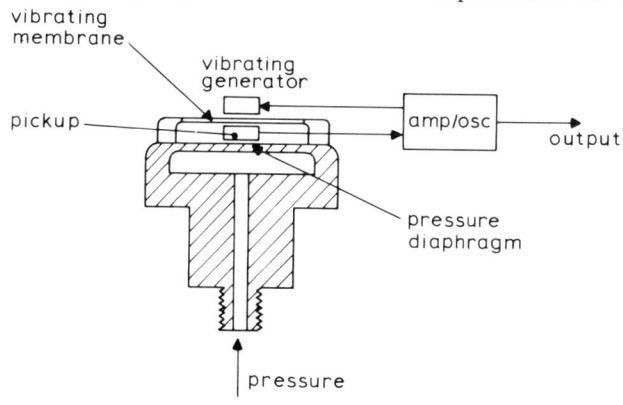

Fig. 3.9 Pressure transducer

A vibrating member attached to two support brackets on the diaphragm is located between a pick-up element and a vibration generator.

Vibration of the vibrating membrane is maintained in the usual manner through an oscillator-amplifier and an electromagnetic vibration generator. Pressure under the diaphragm causes distortion which alters the angle of the support brackets and tensions the vibrating membrane, thus increasing the frequency. The change in frequency of vibration depends on the stiffness of the diaphragm, including the restraints at its circumference, the nature of the stiffness of the support brackets and the vibrating member itself.

Results of two types of transducers, one measuring up to 50 atmospheres and another up to 100 atmospheres, have been found to have an accuracy of approximately 0·1%. Temperature coefficients and dynamic performance are not mentioned in the reference. Suitable design and selection of materials can reduce temperature coefficients to very small values.

3.3 Vibrating cylinder transducers

This type of transducer has been brought to an advanced state of development* and has been adapted for a range of measurements. The essential common feature is the use of the natural frequency of vibration of a cylinder. This natural frequency, similar in principle to the ring produced by a bell, is a function of the physcial properties of the tube or cylinder. It depends on the shape and size, the elasticity of the material, the stresses induced, the mass and also on the distribution of the mass which vibrates.

The parameter to be measured is used to modify one of the properties of the tube, and the change in natural frequency then represents the magnitude of the parameter to be measured. In the majority of cases, the relationship between parameter changes and frequency changes is nonlinear. Primarily, the cylinders can be equated to a mechanical spring-mass-damper system in which the spring rate is a function of the dimension of the cylinder and the induced stress. The natural frequency is then given by the nonlinear relationship

$$\omega_n = \sqrt{\frac{k}{m}} \quad \text{rad/s} \tag{3.7}$$

where k = effective spring stiffness

m = effective vibrating mass

* by Solartron Ltd.

This will be modified slightly by the damping in the system which arises from hysteresis in the stress/strain relationship, when vibrating, and also from the viscosity of the air or gas in contact with the mass. In transducers, calibration is necessary to relate frequency to the parameter being measured.

In the vibration cylinders, changes in temperature affect the physical properties of the cylinders, and it is necessary to use a special nickel-iron alloy. By careful manufacture and the introduction of cold work-

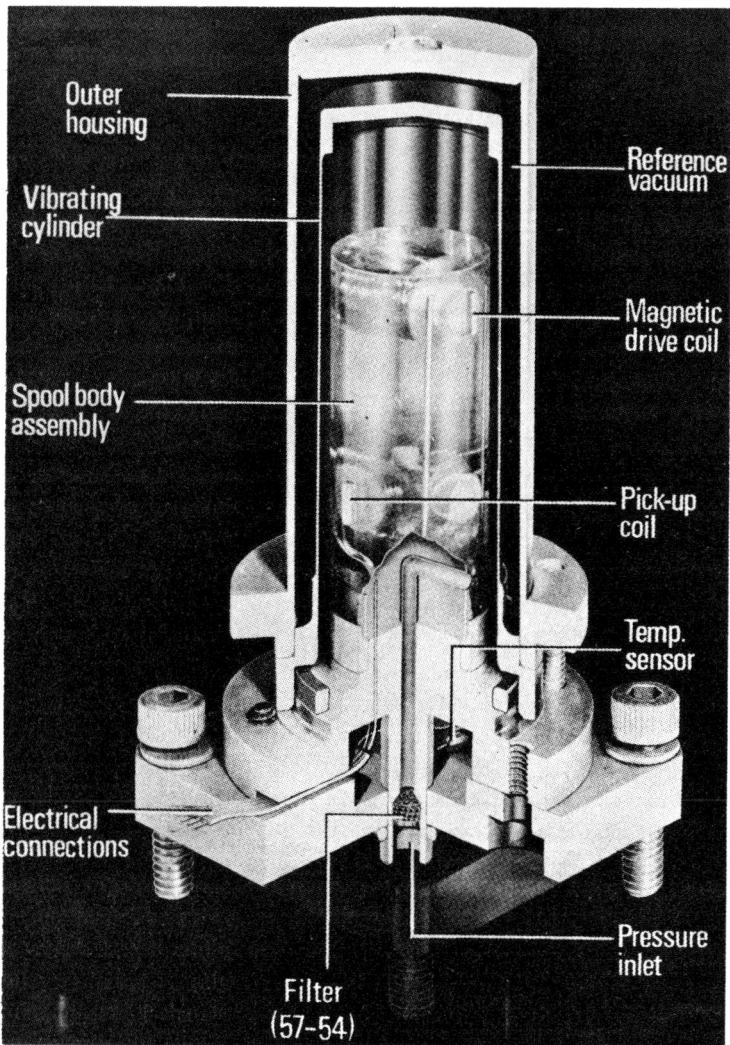

Fig. 3.10 Vibrating cylinder pressure transducer

ing and subsequent heat treatment, a material is produced in which changes of elasticity with temperature have an equal and opposite effect to the changes due to dimensional variations.

Two types of gas transducer have been produced: a pressure transducer in which a pressure change causes a change of stress and hence a change in frequency; and a density transducer in which the density of the gas adjacent to the cylinder walls is effectively part of the vibrating mass and, therefore, changes in gas density gives rise to a change in mass and hence the natural frequency. The unwanted effect of pressure changes in a density transducer can be minimised by ensuring that both the inside and outside of the cylinder are subject to the same gas pressure. However, the density effect in the pressure transducer cannot be entirely eliminated and other means of correction for density changes must be used.

The cylinder is vibrated in the required mode by an electromagnetic system consisting of a magnetic drive solenoid and a pick-up coil. These are located inside the vibrating cylinder at rightangles to each other, to prevent direct coupling as shown in Fig. 3.10. Each consists of a magnetised pole piece and a coil. The pick-up coil feeds into an amplifier, the output of which drives the drive solenoid and also provides the transducer output. Thus, the cylinder is maintained in a state of oscillation, the amplifier providing just enough energy to overcome the inherent damping in the cylinder.

Gas pressure transducer

The construction of a pressure transducer is shown in Fig. 3.10. The vibrating cylinder itself is typically 45 mm long; 19 mm diameter and a wall thickness of 0·075 mm. The gas whose pressure is to be measured, is passed to the inside of the cylinder while the outside is maintained at constant fixed reference, often a vacuum, or, alternatively, to a second source of gas pressure, if differential measurement is required. By using two standard cylinder sizes, it is possible to manufacture transducers having full-scale ranges of 1 to 700 atmospheres. These cylinders give rise to a 20% frequency change over their full working range with a nominal frequency of 5 kHz to about 16 kHz, depending on the pressure range.

The shape of the cylinder, i.e. stiff ends and thin flexible walls limit the modes of vibration to those shown in Fig. 3.11. The mode excited and maintained by the electromagnetic and amplifier system depends primarily on the position of the two coils, and the mode chosen depends on the pressure range to be measured.

Although the relationship between natural frequency and the stress induced by differential pressure across the cylinder walls can be theoretically established, calibration tests show that a typical pressure/fre-

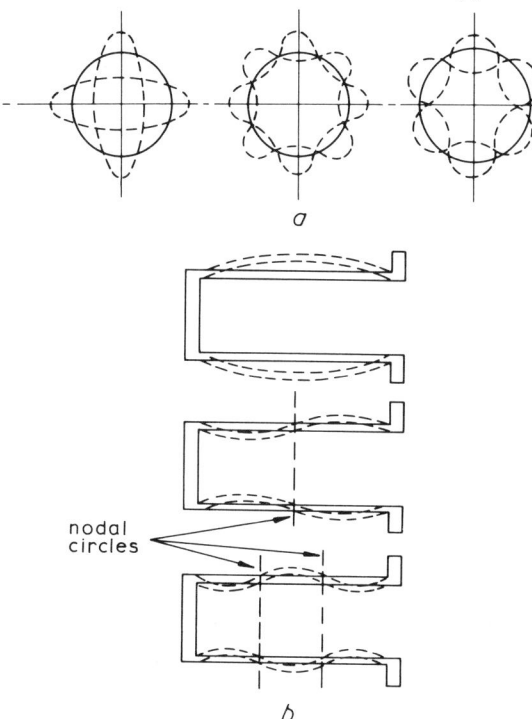

Fig. 3.11 Vibrational modes in cylinders
  a Longitudinal modes
  b Meridinal modes

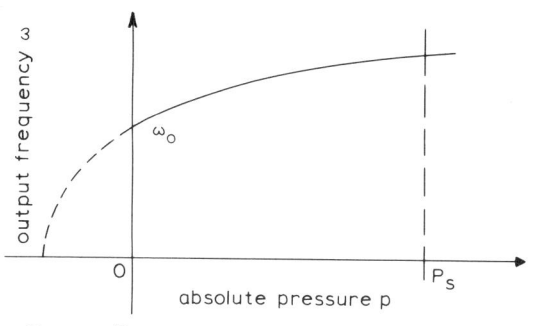

Fig. 3.12 Pressure/frequency relationship
  Calibration at reference temperature
  P_s = full-scale pressure
  $\omega \hat{=} 1 \cdot 2\omega_0$

quency relationship, at a constant temperature, follows a smooth curve, as shown in Fig. 3.12, and has the form

$$P = k_0 + k_1\omega + k_2\omega^2 + k_3\omega^3 \tag{3.8}$$

where P = internal air pressure (vacuum reference)

k_0, k_1 etc.= calibration constants

ω = natural frequency

This relates closely to the physical relationship between frequency and pressure determined analytically.

The nonlinear relationship requires that a correction is included in the measurement system. The system must also take into consideration the change in frequency with temperature which is caused both by physical changes in the material of the cylinder and also its sensitivity to variations in the density of the gas, which also varies with temperature. The temperature coefficient of the cylinder itself is stable to within a few parts per million and may therefore be considered zero for most purposes.

The variation in gas density, due to pressure, changes in part of the basic calibration curve given by eqn. 3.8. It is the variation in gas density with temperature that accounts for the temperature coefficient of the transducer. This variation is given by

$$\Delta f = \frac{k}{R} P \left(\frac{1}{T_1} - \frac{1}{T_2}\right) \tag{3.9}$$

k, R = constants

Δf = change in frequencies due to change in temperature from T_1 to T_2

P = pressure

T_1, T_2 = temperature °K

For air, the total variation in frequency is less than 2% of the mean frequency over a temperature range of $-55°C$ to $+125°C$ and therefore correction for temperature is necessary only where high accuracy is required. The transducers normally have a resistor or diode temperature sensor located in the base from which a resistance or voltage measurement can be made to determine the gas temperature in the cylinder.

The transducer main output will be a train of pulses whose frequency of the cylinder and hence the internal gas pressure. In the usual form of operation the transducer pulses are used to gate pulses from a 10 MHz clock into the counter to provide a measure of the periodic time and hence frequency. Greater resolution can be obtained by accumulating a count over a number of transducer cycles. This also has the advantage that the influence of noise, caused by vibration or electromagnetic changes is minimised. However, the sample rate is naturally reduced, which could raise difficulties in some installations.

Consider a transducer with a nominal natural frequency f Hz and a clock rate F Hz. If a sample is to be taken over N cycles at f Hz, then the sample time is given by:

$$\text{Sample time} = \frac{N}{f} \text{ seconds} \tag{3.10}$$

$$\text{counter resolution} = 1 \text{ in } \left(F \frac{N}{f}\right) \tag{3.11}$$

Over the full pressure range, the frequency changes by only 20%, therefore the resolution on the actual pressure range is given by:

$$\text{measurement resolution} = 1 \text{ in } \left(0{\cdot}2 \times F \frac{N}{f}\right) \tag{3.12}$$

For example, consider a transducer with a nominal frequency of 5 kHz, having a clock of 10 MHz and a measurement sample over 40 periods:

$F = 10$ MHz

$f = 5$ kHz

$N = 40$

then sample time $= \dfrac{40}{5 \times 10^3} = 8$ ms

$$\text{measurement resolution} = 1 \text{ in } \left(0{\cdot}2 \times 10 \times 10^6 \times \frac{40}{5 \times 10^3}\right)$$

$$= 1 \text{ in } 16\,000$$

This gives rise to a 14 bit digital output. Reducing the sample time to, say, 0·1 ms, reduces the measurement resolution to 1 in 200 (only 8 bits).

Where accuracy is important the actual output of the transducer will have to be modified to accommodate both the nonlinear pressure/frequency relationship and the correction which may be necessary to compensate for the temperature. Where the transducer output is connected directly to a digital computer, it may be possible to store a number of points from the pressure/frequency calibration curve and to interpolate linearly between these points for all other values of measured frequency of the transducer. Alternatively, the coefficients of the calibration curve equation may be determined and recorded in the computer and a separate calculation made using eqn. 3.8. Similarly, the compensation for temperature can be easily computed by setting the appropriate constants for eqn. 3.9 into the computer. By measuring the actual temperature from the transducer, the computer can be programmed to make the appropriate correction to the final digital value of the output.

As an alternative to supplying the computer with the transducer characteristics as part of its program, a preprogrammed read-only memory (ROM) can be used which can be attached directly to the transducer. This memory carries all the information regarding the coefficients and temperature scaling relating to that particular transducer. The computer can then scan this memory to assess the data, and refresh this data when required by rescanning. In this way, replacing the transducer with its ROM, automatically changes all the coefficients and constants. If the cylinder frequency is used as the output without any correction techniques, maximum errors in the order of ±2·5% over the pressure range can be expected, if a linear relationship is assumed.

The manufacturers also have available a frequency/analog conversion unit which contains the necessary linearisation and temperature scaling. The overall system then has an accuracy of 0·1% full scale.

Gas density transducer

As mentioned above, the presence of gas adjacent to the walls of a vibrating cylinder adds to the mass. The frequency therefore changes according to the density of the gas, providing that the gas flow along the cylinder walls is maintained within reasonable limits, and the gas pressure on each side of the cylinder is the same.

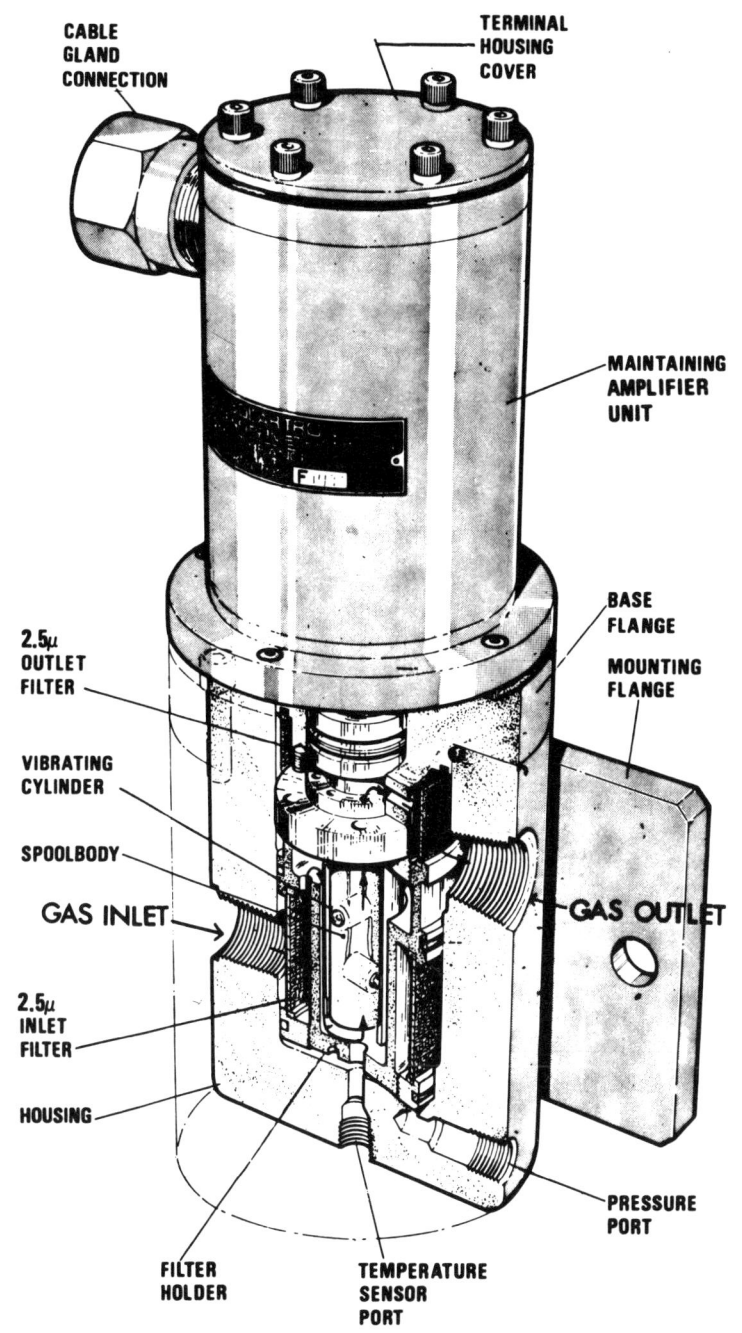

Fig. 3.13 Gas-density transducer

Fig. 3.13 shows the contruction of the transducer* described by Potter 1969. The gas flow is arranged to be in contact with both sides of the cylinder walls and as there is no pressure difference, any change in frequency must be due to changes in the gas density. An increase in density increases the effective mass and lowers the natural frequency of vibration. Oscillation is maintained in the same manner as in the pressure transducers.

A low temperature coefficient is achieved by machining the cylinder sensing element from the special alloy described before. Over the temperature range $-10°$ to $85°C$ (the maximum temperature range), the output variation is about 0·005%. Changes in density due to temperature changes, which cause an error in the pressure transducer, are not relevant in the gas-density transducer as it changes in density, whatever the reason, that are being measured.

As for the pressure transducer, the mechanical design gives rise to a high immunity from external mechanical vibrations and to other possible errors which may arise from creep, drift and hysteresis.

The cylinders used are also similar in size to those used in pressure transducers, with wall thicknesses of 0·05 mm to 0·15 mm depending on the density range to be measured. Again, similar to pressure transducers, a 20% frequency variation is taken to correspond to approximately full-scale density range and currently two ranges are available.

density	frequency range
0 - 60 kg/m³	4·9 - 3·9 kHz
0 - 120 kg/m³	7·04 - 5·64 kHz

The calibration curves for these transducers show that the relationship between the output frequency of vibration and the gas density follows the relationship:

$$\rho = k_0 + \frac{k_1}{\omega} + \frac{k_2}{\omega^2} \quad (3.13)$$

where ρ = gas density

ω = cylinder frequency

k_0, k_1, k_2 = constants for each transducer

*produced by Solartron Ltd.

When operating a particular transducer over a narrow density range, a linear relationship can often be assumed which allows the actual frequency to be directly interpreted as a measure of density. For maximum accuracies, i.e. 0·1% over the full density range, the nonlinear eqn.3.13 must be used, in which case some correction must be applied to the measured frequency, as described for pressure transducers.

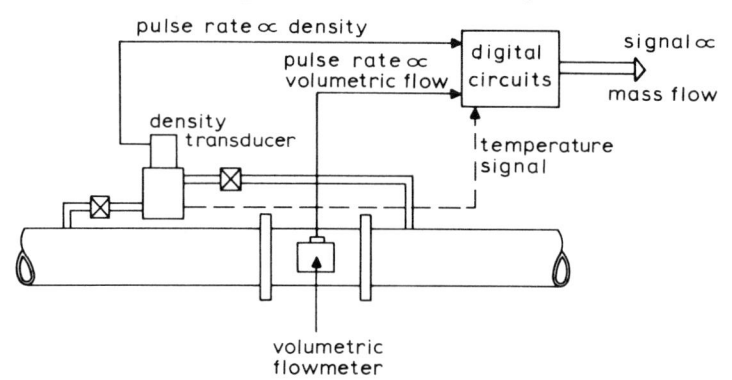

Fig. 3.14 Installation for density and mass flow measurement

Within limits the transducers are independent of flow rate, and in practice the flow is adjusted by valves as shown in Fig. 3.14. The filter in the transducer does create a small pressure differential and at high flow rates through the transducer there could be a difference in gas pressure at the cylinder compared to the main pipeline and therefore an additional small measurement error might occur. The magnitude of the pressure differential can, however, be used as a measure of the condition of the filter.

Mass flow measurement

Density transducers play an important part in accurate mass flow measurement which is of particular importance where large scale metering is required for accurate accountancy purposes. Two standard methods of flow measurement which can be used in conjunction with the density meter are orifice plates and venturi tubes. In each case the basic flow relationships are

$$\text{volumetric flow} = k\sqrt{h/\rho} \tag{3.14}$$

$$\text{mass flow} = k\sqrt{h\rho} \tag{3.15}$$

k = meter constant

h = differential head pressure across device

ρ = gas density

The accurate measurement of density with the density transducer enables determination of mass flow to a greater accuracy than any of the more conventional techniques.

When turbine flowmeters are used for volumetric flow, it is only

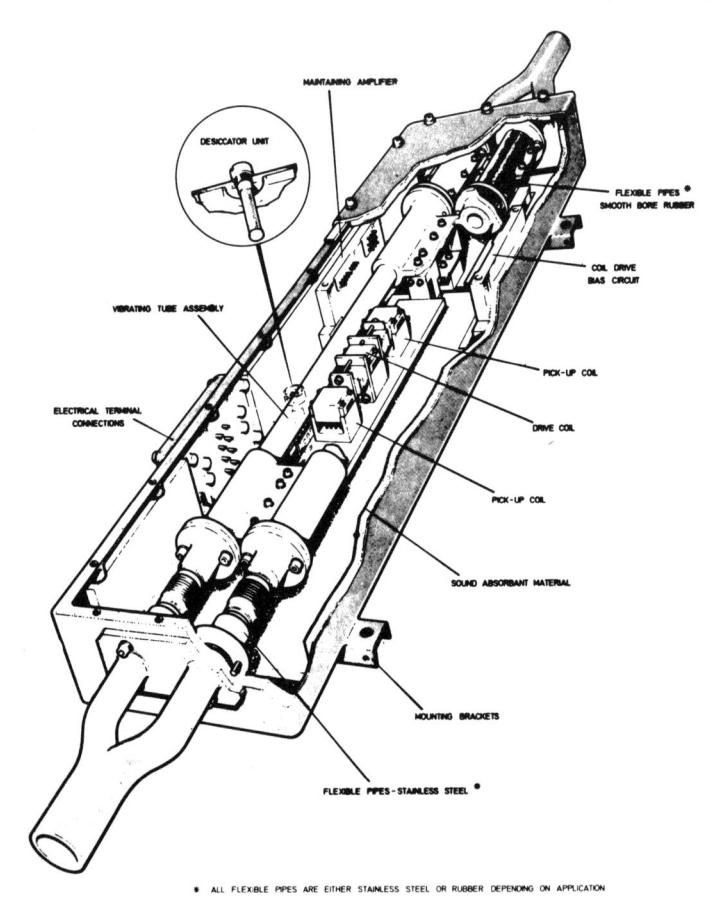

Fig. 3.15 Liquid-density transducer

necessary to digitally multiply the turbine pulse rate and the frequency of the density transducer to obtain a measure of mass flow. Digital techniques can also be used with orifice plates if a digital pressure transducer is used with the density transducer.

The usual arrangements for mass flow measurements are shown in Fig. 3.14. The gas input and output of the density meter are spaced each side of the turbine flowmeter which creates a pressure differential to induce flow through the density meter. The digital circuits will contain the necessary clock and linearisation control circuits as well as the mass flow computation circuits.

Liquid density transducer

This transducer * differs from those previously described in that the induced vibration is a mode involving the whole length of the tubes relative to their fixed ends, rather than a distortion of the tube cross section. From Fig. 3.15 it will be seen that the liquid flows through two parallel tubes in the transducer. The ends of these tubes are secured together and to the rigid case leaving a free central span, the pickup and drive coils being located between the tubes along this freespan. Flexible connections to the ends of the tubes prevent thermal expansion and stresses from external plant pipework from affecting the tube vibrations.

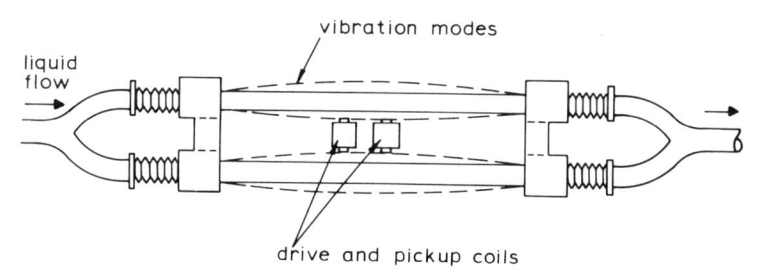

Fig. 3.16 Vibration modes liquid-density transducer

The tubes vibrate in a lateral mode as shown in Fig. 3.16, and the natural frequency depends on the mechanical stiffness and on the mass per unit length of the tube and its contents. The relationship between the liquid density and output frequency follows the relationship

*developed by Solartron Ltd.

$$\rho = k\left(\frac{\omega_0}{\omega}\right) - 1 \tag{3.16}$$

where

ρ = liquid density

ω_0 = frequency at zero density

ω = frequency at density being measured

k = constant

As in the vibrating cylinder transducers, the dimensions of the tubes are chosen so that there is about 20% change in frequency for a density change from 0 to 1 kg/m^3, with a maximum frequency of about 1350 Hz with only air in the tube. Calibration is done with liquids of known densities, any nonlinearity is corrected in the measuring system if the maximum accuracy of 0·01% is required over the whole range.

The calibration remains accurate to within 0·3 x 10^{-4} kg/m^3 over a range of flowrates from zero to 4500 litres/h. Larger flow rates can be accommodated by a simple by-pass arrangement. It is recommended by the manufacturers that the transducer is installed vertically with the flow of liquid from top to bottom. This minimises the erosion and the possibility of deposition on the tube walls which gives rise to incorrect readings, particularly at the lower rates of flow when the solids tend to sink and rest against the tube walls if in a horizontal position. It also allows bubbles to float away and thus prevent incorrect readings if gas bubbles are entrained in the liquid within the tubes.

There will be a small pressure coefficient which is approximately 0·6 x 10^{-5} kg/m^3 per kN/m^{-2} up to the maximum operating pressure, which in some models is 7000 kN/m^2. This deviation can, however, be compensated in the measuring system.

Measurement and reading is similar to those used in other transducers of this type. Direct computer measurement of the frequency allows all the linearisation and compensation to be programmed and also some degree of self checking, as the frequency must be in the range 1050 to 1350 Hz. Frequencies outside this range indicate some malfunctioning of the transducer or the transmission system.

Furthermore, an additional input from a volumetric flow meter, such as a turbine-type transducer, enables an accurate computation of liquid mass flow in a manner similar to gas mass flow measurement described earlier.

The vibrating cylinder transducers, all the types described above, have proved to be remarkably stable over long periods and have provided the processing and aviation industries with a range of reliable and rugged transducers which are able to operate accurately over a wide range of temperatures. The principle is also capable of further development for the measurement of other physical properties and mechanical forces and, no doubt, for many more specialised applications.

3.4 References

APPLICATIONS ENGINEERING GROUP OF HEWLETT-PACKARD LTD' (1965): 'A quartz thermometer', *Instrument Practice,* **19**, (10)
BELYAEV, M.F., et al. (1965): 'Vibration-frequency pressure trans--ducer', *Instrument Construction,* pp. 10-13
HALFORD, R.J. (1964): 'Pressure measurement using vibrating cylinder pressure transducer', *Instrument Practice,* **18**, (8)
HAMMOND, D.L., and BENJAMINSON, A. (1968): 'The crystal resonator—a digital transducer', *IEEE Spectrum*, pp. 53-58
LOVBORG, L. (1965): 'A linear temperature-to-frequency converter', *J. Sci. Instrum.,* **42**, pp. 611-614
POTTER, P.N. (1969): 'Frequency domain transducers and their applications', *Instrument Practice,* **23**, (12)
VOUTAS, A.M. (1963): 'Twisted beam transducer', *J. AIAA,* **1**, (4)
WYMAN, P.R. (1973): 'A new force-to-frequency transducer' *in* 'Digital Instrumentation'. IEE Conf. Publ. 106, pp. 117-123

Chapter 4

Digital linear transducers

4.1 Using rotary encoders

In this chapter the word linear is used to denote translation motion as against its other meaning which implies a mathematical relationship between two variables.

Many of the transducers used for measuring linear displacement and motion are adaptations of the techniques used for rotational displacement and velocity, described in earlier chapters. The simplest arrangement is the conversion of linear motion to rotary motion through a rack and pinion or a nut and screw device. The accuracy of these systems depends on the linearity of the mechanical linear-to-rotary converter.

A specially designed system using a shaft encoder to measure linear displacement has been marketed* which depends for its accuracy on a preloaded nut which operates on a ballscrew shaft (Fig. 4.1). The application is intended for use on machine tools and similar equipment. Rotation of the shaft will cause linear displacement of the nut to which the surface to be moved is attached. The shaft carries a standard in-

*by Vatric Ltd.

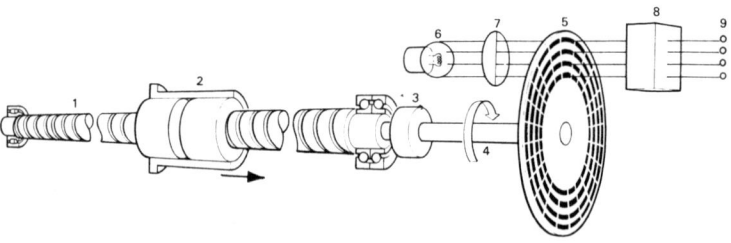

Fig. 4.1 Linear displacement transducer
    1 Ballscrew   6 Lamp
    2 Preloaded nut assembly   7 Lens
    3 Flexible coupling   8 Photo cells
    4 Encoder shaft   9 Cell outputs
    5 Code disc

cremental encoder disc which measures the shaft position by counting pulses developed by the encoder output.

Four separate channels on the disc are used to identify the direction of rotation, as explained in Chapter 2. The pulse counting system could also include a datum setting control which can set the digital output to zero when the nut is in the datum position. Rotation of the shaft and hence movement of the nut will be recorded by the counter as a linear displacement, positive or negative, from the datum position. The system has a high degree of accuracy, approximately 0·003 mm over a traverse of 1 m, due to the high precision achieved in the ballscrew assembly, and the special shaft bearings which prevent any longitudinal displacement. An overall resolution of 0·01 mm is achieved.

An alternative arrangement of the assembly described above is to mount the ballscrew in bearings parallel to the moving member whose displacement is to be measured. An attachment is made from the moving member to the nut of the screwshaft assembly. Any movement of the nut causes the screwshaft to rotate, thereby rotating the encoder disc.

A simpler arrangement makes use of an accurately-machined rack

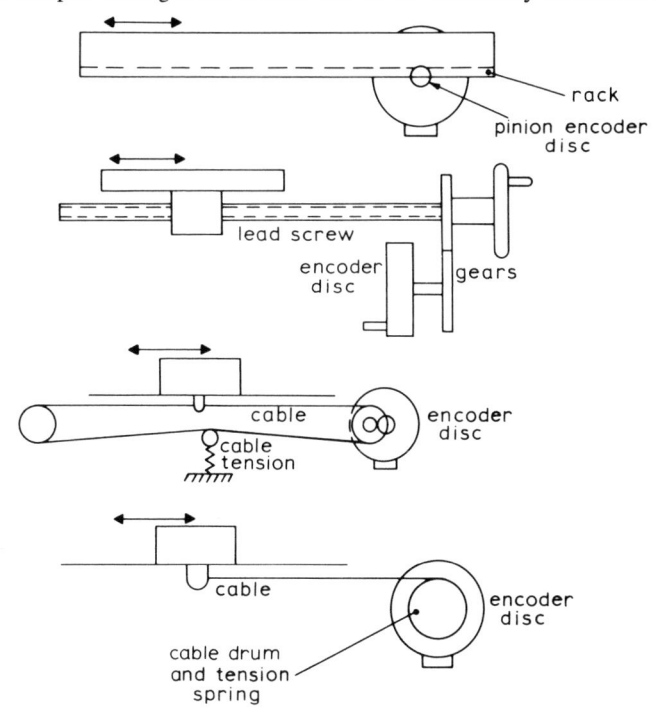

Fig. 4.2 Linear/rotary conversion

attached to the moving member which engages with a pinion rotating a shaft encoder. The accuracy will depend primarily on the rack and pinion and the mechanical assembly of the system. Similar conditions will apply if the machine leadscrew is used and the encoder attached directly to the leadscrew or driven through a gear train from the leadscrew. More accurate drive to the digitiser disc can often be made by friction drives, either directly by a pulley bearing on a flat surface or by a cable as shown in Fig. 4.2.

When absolute encoders are used in the above installations, the resolution of the disc will limit the linear traverse that can be measured for a single rotation of the disc. In this case gearing will be necessary to provide drives to two or more discs to give fine and coarse drives to separate encoders.

With all the above systems, the precision of the mechanical components and assemblies is of fundamental importance if a high degree of accuracy is to be maintained. The precautions outlined in Chapter 2 for mounting encoder discs are added to by the additional demands made on the mechanical assemblies which translate the linear motion to rotary motion, if the resolution and accuracy available from the shaft encoder are to apply to the measurement of the linear motion.

4.2 Linear encoders

To avoid the difficulties associated with the use of rotary encoders, direct absolute and incremental linear encoders have been developed. These usually involve the use of a long scale and a scanning head which moves along the scale, sensing and measuring any relative displacement. The advantages are that the coarser tolerances and wear associated with the transfer device (linear-to-rotary) do not arise and have no effect on the positioning accuracy of the transducer and its output.

The scale is, in effect, a straightened or linear version of a shaft encoder, consisting of rows or tracks of opaque and transparent areas, coded in binary or BCD as required. It is also possible to use contact linear encoders but the difficulty of completely enclosing the contact strip generally creates difficulties with contamination. Some optical systems have a linear scale consisting simply of a finely divided optical grating producing alternating transparent and opaque areas. The displacement is measured by the incremental technique, using light cell sensors, as with shaft encoders. More than one track is usually necessary to identify direction of movement but any technique which provides two square wave outputs, displaced by $90°$ can be used. The

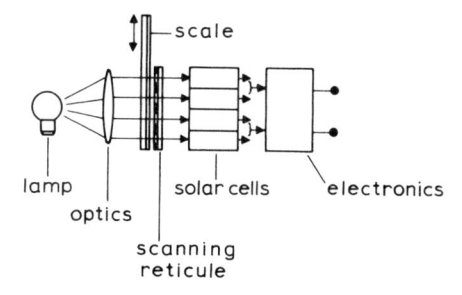

Fig. 4.3 Transmitted-light linear encoder

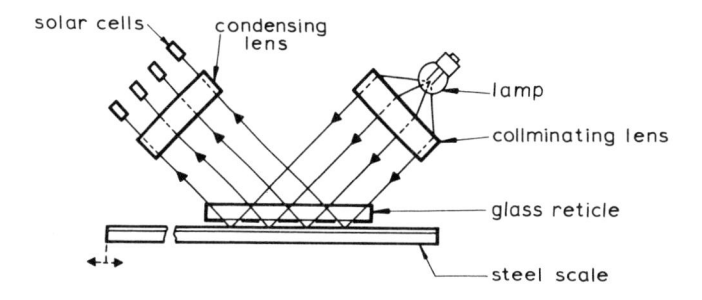

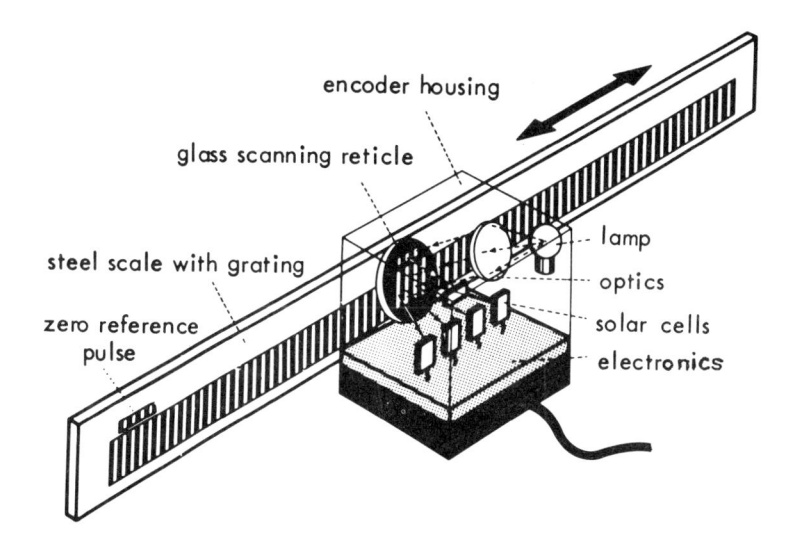

Fig. 4.4 Reflected-light linear encoder

counting circuits used are similar to those described in Chapter 2. The scanning head contains a scanning reticle with windows positioned adjacent to each track. Two techniques are in use for detecting relative displacement between the scale and the scanning head: these are the transmitted-light method and the reflected-light method. The transmitted-light method, shown in Fig. 4.3, is similar to that used in angular digital encoders.

The reflected light method is one which has particular application to machine tool slides and is usually limited to the use of the incremental method of measurement. The grating is etched on a stainless steel tape which is cemented into a stainless steel carrier attached to the sliding surface. The illumination from the lamp passes through the reticle onto the scale, and reflected light passes again through the reticle gaps onto sensing cells, as shown in Fig. 4.4.

Displacement of the scale relative to the scanning head will cause light fluctuations on the cells which generate outputs of a sinusoidal character, which can be shaped for counting purposes. Maximum light intensity on the cells is reached when the transparent gaps in the scanning reticle are positioned directly above the reflecting lines of the steel scale.

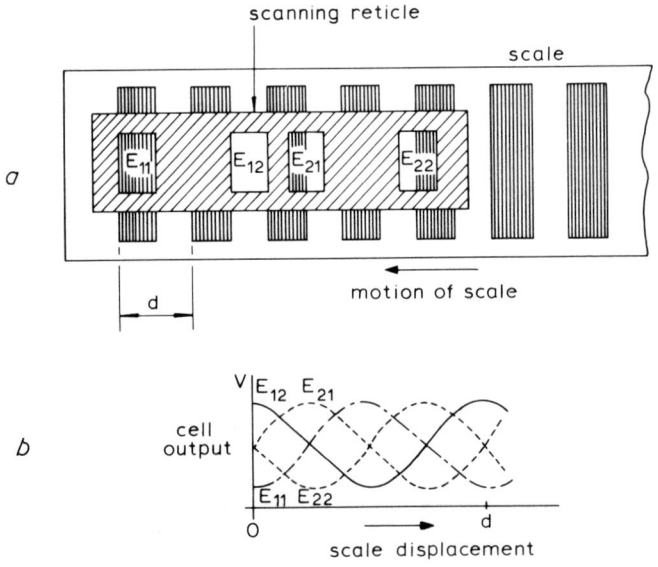

Fig. 4.5 Four-phase cell output
    a   Four-phase reticle
    b   Cell outputs

Incremental encoder

The most popular systems currently available use the incremental technique with a single track graticle and either the transmitted-light method or the reflected-light method. The scanning-head graticle, or reticle, has four separate grating windows positioned so that the outputs of the cells are a 0·25 grating-period shift. Therefore, the quasi-sinusoidal signals of the cells are 90° phase shifted relative to each other (Fig. 4.5) due to the offset scanning windows. The signal element E_{21} shows a 90° phase shift as compared to E_{11}. The signal E_{12} and E_{22} are 180° phase shifted compared to the signal E_{11} and E_{21}.

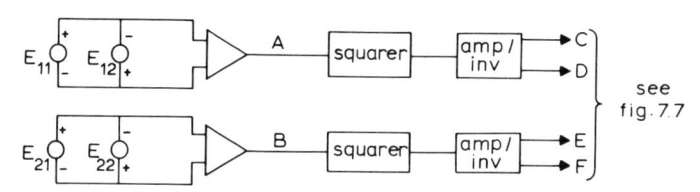

Fig. 4.6 Cell output pulse shaping

The pairs of cells, (E_{11}/E_{12}) and (E_{21}/E_{22}) are connected into a push-pull arrangement as shown in Fig. 4.6. Subsequent electronic circuits, Schmitt trigger circuits, amplifiers, inverters etc. convert the outputs into a square wave and its inversion, the outputs of the E_{11}/E_{12} circuits having a 90° phase shift with respect to the outputs of the E_{21}/E_{22} circuits.

The resulting wave forms are shown in Fig. 4.7. The upper curves are the signals developed by the cells. Subsequent diagrams show the relationship between the various square waves developed by the electronic circuits. One-shot circuits triggered by the positive-going edge of these square waves, enables the generation of counting pulses either one pulse for pattern period, or two or four pulses per pattern period. In this way a grating having a pitch of 0·01 mm could have a resolution of 0·0025 mm.

The maximum counting speed of the bidirectional counter will limit the maximum speed of the transducer. For example, if the resolution is X mm and the maximum counting frequency is Y Hz, then the maximum velocity is XY mm/s.

In addition to the measurement grating, some systems include a second short additional track to provide a fixed absolute reference mark. This can be seen on the track shown in Fig. 4.4 and is marked as a zero reference pulse. This 'zero' marking pulse makes the incremental coder into a quasi-absolute system and is used for re-establishing a reference position after interruptions. The generation of the two sine waves at 90° displacement makes it possible to use an interpolator as discussed later in this chapter to generate a higher degree of resolution than is generated even by the four count pulses per graticle pitch.

One of the advantages of any incremental method is that measuring the time interval between pulses gives an indication of velocity which can be an advantage in some control situations.

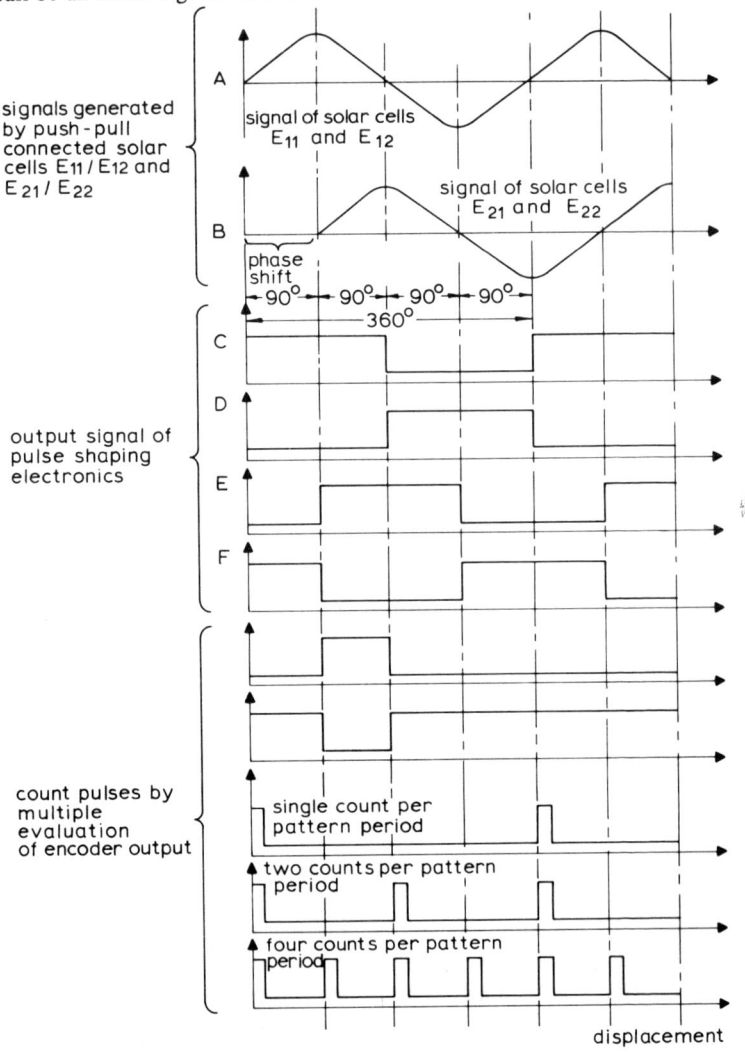

Fig. 4.7 Scanning head output waveforms

Mechanical details

The above principles have been used to develop a range of transducers for a wide variety of applications. Completely sealed units containing their own guide bearings for the push rods are available and resemble large linear potentiometers. Units are currently available in this form and are capable of measuring up to 0·5 m with resolutions of 0·0005 mm, with overall system accuracies up to 0·001 mm. These units use the transmitted-light method, but, being completely sealed, they are immune to errors in reading which can arise from contamination of the scales.

Other types of assembly using the transmitted-light method have open exposed scales with a separate scanning head containing the graticle, light cells and pulse-forming electronics. These units are available for displacements up to 1·2 m with resolutions and accuracies similar to the sealed units. The scales must be mounted to leave a clear space each side for the scanning head and this could at times lead to difficulties in assembly. Further, as already mentioned, the exposed scale can lead to contamination or damage which can cause miscounting and inaccurate measurements. Contamination detecting systems are available, however, similar in principle to the detection of lamp deterioration, discussed in Chapter 2.

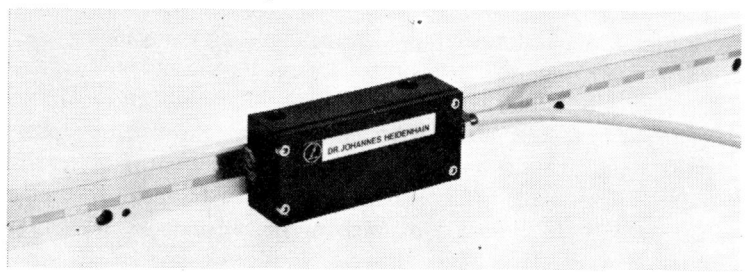

Fig. 4.8 Reflecting-light linear transducer

The reflecting-light method is generally more robust as the scale can be secured over its entire length, permitting longer measurements. Fig. 4.8 shows a typical assembly*. This system depends on the guideways of the machine to which it is attached for accurate movement of the scanning head over the scale. These systems are available in lengths up to 3 m with resolutions up to 0·002 mm and overall accuracies up to 0·003 mm.

*supplied by Heidenhain (GB) Ltd

Inductosyns

The principle of the synchro (Chapter 6) has been adapted for the measurement of linear displacement. Inductosyns are effectively synchros which have been flattened so that the stator is formed into the long flat scale, consisting of a pattern of conductors arranged as shown in Figs. 4.9 and 4.10. The scanning head contains two smaller similar patterns which relate electrically to the conductors on the scale in a similar relationship to the two rotor coils and the stator in a synchro resolver.

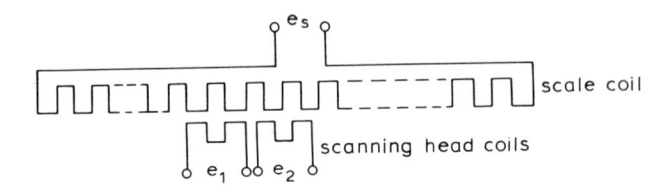

Fig. 4.9 Inductosyn coils

The pitch of the conductors is maitained accurately over the whole length of the scale and also along the coils of the scanning head. This pitch is usually about 2 mm. The two coils in the scanning head are displaced relative to each other by 0·25 of the pitch. The units behave like a flattened transformer in which the 'coils' on the scanning head are the primaries and the 'coil' on the scale is the secondary. Two sine voltages, usually at 400 Hz, and one shifted 90° relative to the other, are supplied to the scanning head 'coils'. Providing the distance separating the slides and the scanning head remains constant, a voltage is introduced in the scale 'coil'. This voltage is the sum of the sinusoidal voltage at 400 Hz produced by each of the separate 'coils' in the scanning head.

If the scanning head input voltages are

$$e_1 = e \sin \omega t \tag{4.1}$$

$$e_2 = e \cos \omega t \tag{4.2}$$

ω = supply frequency

Then the induced voltage in the scale 'coil' will be

$$e_s = e_x \sin(\omega t + gs) \tag{4.3}$$

where g represents the grating pitch or period and s the displacement of the first scanning 'coil' e_1 within one grating pitch or period. For example, e_s will go through one complete cycle when the displacement is equal to one grating pitch of 2 mm. Thus, a 360° or 2π phase change occurs for a displacement of 2 mm. Therefore, for a displacement of 0·3 mm, $gs = 0\cdot3/2 \times 2\pi = 0\cdot3\pi$ radians.

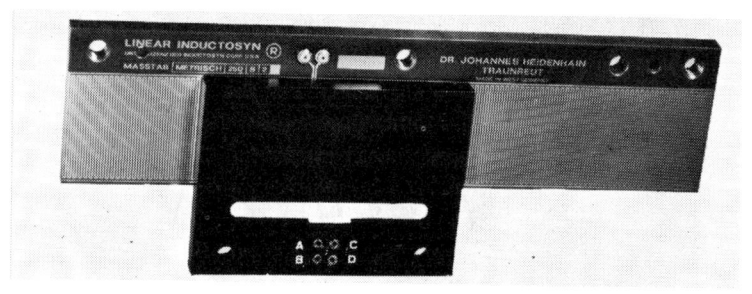

Fig. 4.10 Commercial Inductosyn

Similarly, the amplitude will vary with displacement and also with any unintentional alteration in the gap separating the coils.

The scale 'coil' output can then be decoded by identifying the change in amplitude or, more usually, by determining the phase changes using the methods adopted in synchro technology. Digital determination of the resulting displacement measurement can then follow the methods described in Chapter 6 on synchro-digital resolvers.

Commercial units currently available* have a scale 250 mm long and can have accuracies up to 0·001 mm. Longer lengths are also available and angular inductosyns have been produced to determine relative angular displacements.

4.3 Moiré fringe techniques

Moiré fringes are the dark patterns produced when two identical gratings are superimposed but are slightly out of alignment with each other. A grating is a transparent surface having a large number of ruled lines, evenly spaced and parallel. They were originally produced for use in spectrometers but are now used in a wide range of measuring devices. If two identical gratings are superimposed with the lines parallel, the overall effect to transmitted light will be to block it if the lines on the upper grating are located between the lines of the lower grating, or to

*from Heidenhain (GB) Ltd.

let it pass through if the lines of the two gratings are directly opposite each other.

Thus, if one grating is moved relative to the other in a direction at right angles to the lines, the appearance is of alternating between light and dark. A movement equal to the pitch of the grating produces one complete cycle of light-dark-light.

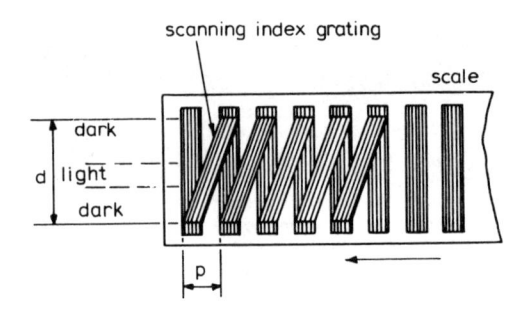

Fig. 4.11 Moiré fringe

However, if the gratings are angled with respect to each other, as shown in Fig. 4.11 where the inclination is equal to the pitch p over a width d of the grating, then similar displacement of one with respect to the other gives rise to Moiré fringes, that is dark strips at right angles to the lines. The fringe moves upwards, for displacement in the direction shown, and moves the distance d for a displacement equal to the pitch p. For displacement in the other direction, the fringe appears to move downward in Fig. 4.11.

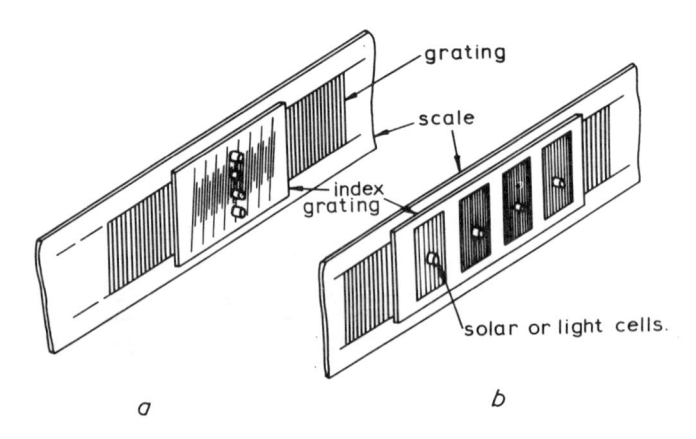

Fig. 4.12 Type of index gratings

The effect of this is to provide a detectable fringe which moves a relatively long distance d for each displacement equal to the pitch.

In application for measurement purposes, there is a long scale consisting of the grating and a separate scanning head that carries a lamp and optics to illuminate four light cells across the width of the second grating, known as the index grating, which is also carried in the scanning head (Fig. 4.12a). In some cases the index grating consists of four separate short lengths of grating, (Fig. 4.12b), which allows the four cells to be placed side by side along the length of the scale, thus permitting a narrow grating to be used on the actual scale. In this system, the lines of the scale and the index gratings are parallel and the cell detects the light/dark pattern over the whole area of the grating segment, by the method described earlier. The four segments are displaced such that each produce signals 90° out of phase with the next.

Whatever the arrangement, the aperture and optics can be arranged if required, such that each cell produces a near sine wave output as the fringe interrupts the light. Each cell output is displaced by 90° electrically with respect to its neighbour as illustrated in Fig. 4.5b.

Each cell output could be shaped into a square wave and used to generate four counting pulses for a displacement equal to the pitch of the grating.

Gratings having 40 lines/mm are commonplace and can give a resolution of better than 0·005 mm. More usually, the cell outputs are combined in pairs to generate two sine waves with 90° phase difference and these are used to generate the necessary direction indication signal and either 1, 2 or 4 counting pulses per pitch as previously described and illustrated in Fig. 4.7. It is also necessary to provide some datum or zero setting mark on the scale for most industrial purposes.

These units are susceptible to counting errors which may arise from electrical interference or contamination of the grating. The fact that the cells are sensing a fringe which is produced by the cumulative effect of a number of lines on gratings does to some extent reduce the susceptibility to errors due to contamination compared to other counting methods where individual lines or opaque areas are used to determine counting pulses. This averaging effect also minimises inaccuracies in the gratings themselves, although the manufacturing methods used, as in the case of shaft encoder discs, are such that the accuracies achieved are adequate for most industrial and scientific purposes.

Increasing the resolution

With any incremental displacement system it is necessary to develop two sets of pulses differing by 90°. The resolution can be increased from one pulse per cycle to two or four pulses per cycle by techniques already discussed, but a further increase in resolution can be obtained by the use of an interpolator, first discussed by Russell (1966) in a publication from the National Engineering Laboratory. This method can be adopted for use with linear or rotational encoders.

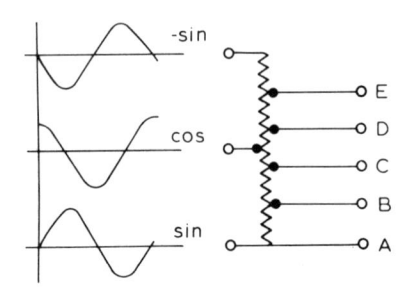

Fig. 4.13 Curve generator or 'interpolator'

The two quasi sine and cosine waves, together with inverted sine, are applied to an interpolator, the simplest design being the resistor network shown in Fig. 4.13. The outputs at a, b, c, d and e produce a pattern of quasi sine waves, differing in phase according to the tapping position on the resistance chain. In practice it is found that the overall performance of the system can be improved by using near triangular waveforms and this can usually be accommodated by design of the mechanical arrangements in the scanning head.

Each separate waveform from the pattern is then used to generate square waves, by the use of level detectors, which differ in phase relationship as shown in Fig. 4.14. By using a relatively simple logic system ten separate pulses can be generated as follows:

$$0 = A\bar{B}$$
$$1 = B\bar{C}$$
$$2 = C\bar{D}$$
$$3 = A\bar{E}$$
$$4 = AE$$
$$5 = \bar{A}B$$
$$6 = \bar{B}C$$
$$7 = \bar{C}D$$
$$8 = \bar{D}E$$
$$9 = \bar{A}\bar{E}$$

The counting pulses are generated by one-shot units triggered by the positive-going edge of each of the ten pulses produced from each cycle.

In this way the outputs from, say, a grating having 40 lines/mm have the resolution increased from 0·025 mm to 0·0025 mm by using an interpolator producing 10 pulses/cycle. Even better resolution can be obtained by using an interpolator which generates more than the ten outputs per cycle.

By using an alternative logic in the decoder it is possible to make the interpolator generate a binary or Gray code output. For example, an interpolator with four outputs (A, B, C and D) per grating division, the

A. $\bar{B}$	0
B. $\bar{C}$	1
C. $\bar{D}$	2
D. $\bar{E}$	3
A. E	4
$\bar{A}$. B	5
$\bar{B}$. C	6
$\bar{C}$. D	7
$\bar{D}$. E	8
$\bar{A}$. $\bar{E}$	9

counting pulses

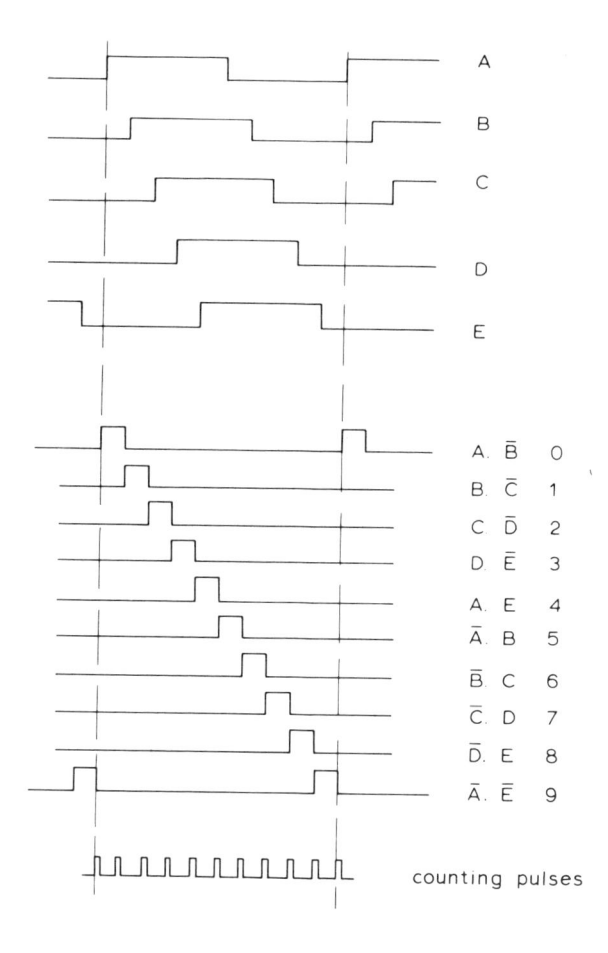

Fig. 4.14 Interpolator logic

positions between each grating division can be identified by a 3 bit binary number derived from the four outputs as follows:

$$000 = A\,B\,\bar{C}\,\bar{D} \qquad 100 = \bar{A}\,B\,C\,D$$
$$001 = A\,B\,\bar{C}\,\bar{D} \qquad 101 = \bar{A}\,B\,C\,D$$
$$010 = A\,B\,C\,\bar{D} \qquad 110 = \bar{A}\,\bar{B}\,\bar{C}\,D$$
$$011 = A\,B\,C\,D \qquad 111 = \bar{A}\,B\,\bar{C}\,\bar{D}$$

Some of the difficulties with these and all systems depending simply on pulse counting is the loss of correct measurement if, power to the system is interrupted, false counts occur due to electrical interference, or if increments are not detected due to contamination on the grating.

Absolute transducer using grating

To overcome some of the difficulties of incremental devices, the National Engineering Laboratory (1966) developed an absolute measuring system that uses an optical grating with shaft encoders. The accuracy of the complete system is determined by the highly accurate optical grating.

The system is unique in the method used to relate the encoder and grating signals without the necessity of a highly accurate shaft-encoder/ grating mechanical relationship. The system uses interpolation of the grating pitch described above to increase the resolution of the shaft encoder and provide some sensing logic. The following description refers mostly to a system used to generate decade signals, but other outputs, i.e. binary of BCD can be used. Each of the ten subdivisions of the grating pitch, generated by the interpolator, is used to provide an absolute measurement between each of the grating increments.

The shaft encoders, attached to leadscrews driving the grating indexing head, must now be related so that the head develops absolute output signals which have a resolution equal to each grating division. Thus, the shaft encoder output, which is an absolute measurement is increased by one decade of absolute measurement from the grating.

Fig. 4.15a shows a relationship between the grating divisions and the encoder resolution which indicates that a precise mechanical relationship is required between the encoder and the grating if ambiguity is to be avoided at the transition points of the encoder.

This requirement is avoided in the NEL system by modifying the relationship to that shown in Fig. 4.15b. The encoder track and the grating division need now only have an accuracy of up to ± half a

Digital linear transducers 93

division of the grating, providing this tolerance is not exceeded over the whole length of the grating. The encoder is interrogated only during the transition of the interpolator output from 0 to 9 or 9 to 0, i.e. at each grating division. Other, coarser, tracks of the encoder can be interrogated in a similar manner.

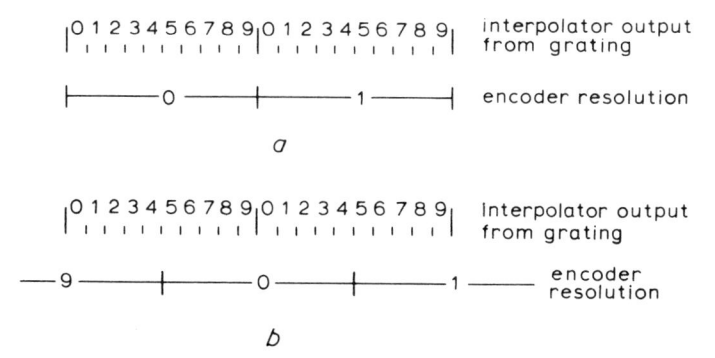

Fig. 4.15 Grating and encoder relationship

However, this system still has three difficulties which must be overcome. These are:

(i) while displacement in one direction, which causes a transition from 9 to 0 of the interpolator output, calls for a number X to be registered when the encoder is strobed, a transition from 0 to 9 requires that $(X-1)$ is registered;

(ii) when a transition does occur, the successive readings from the

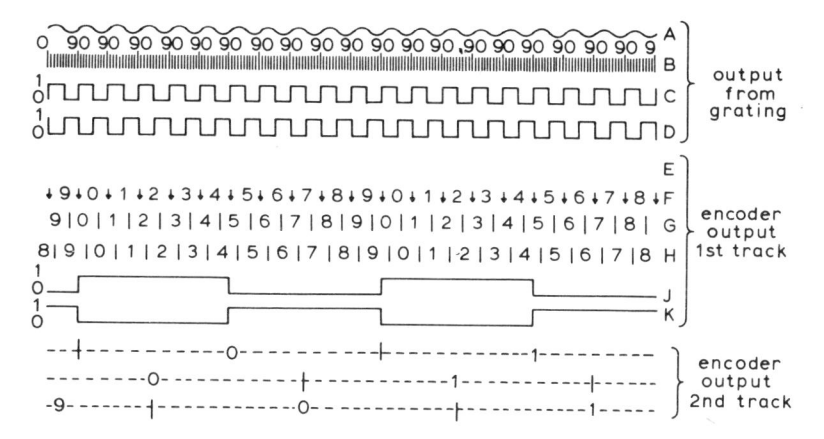

Fig. 4.16 Interrogation of grating-encoder combination

encoder tracks must be stored. Therefore, after an interruption, the stored information may be incorrect until such time as the 0 to 9 or 9 to 0 transition of the track occurs, which calls for all encoder tracks to be interrogated again;

(iii) all encoder tracks must be related to the grating with a maximum tolerance of ±0·5 unit of the grating.

The system described by Russell (1966) which overcomes these difficulties is detailed in Fig. 4.16. The upper four rows are the outputs and the derived outputs from the grating. Row A represents the cycle of output from the grating divisions, e.g. each cycle representing 0·1 mm for a 10 lines/mm grating. Row B represents the ten subdivisions generated by the interpolator, each subdivision representing 0·01 mm. Row C is a 'lead' line which is a square wave at the same rate at the grating output and is a logic '1' for counts of 0 to 4 of the interpolator output and at logic '0' for counts 5 to 9. Row D is a 'lag' line and is the inverse of row C.

Rows E to K represent the signals derived from the shaft encoder which is of conventional form having binary or Gray code output which can, if required, be decoded into decades. Assuming that the latter is

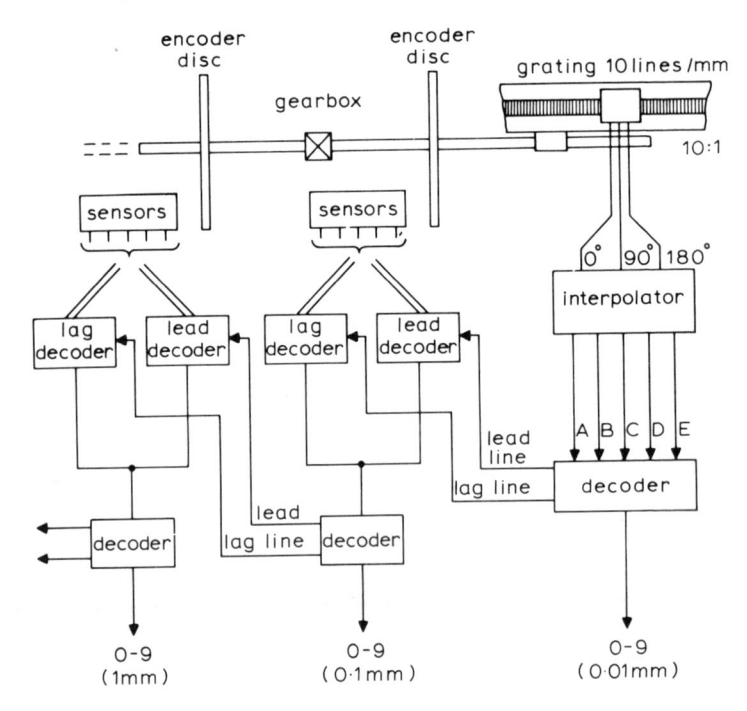

Fig. 4.17 Grating-encoder system

the case, i.e. that the output is in decades, then the tracks on the encoder disc to provide the next decade of output are required to generate two outputs. One provides the ten signals (representing 0 to 9) which lead on the main grating division and the second set which lag on the main grating division. These are shown in rows G and H, respectively. Each of the ten outputs will have to be generated from the shaft encoder outputs.

The least significant digits of the overall transducer output will be made up of the grating interpolator output and one of the two encoder outputs, either the 'lead' output or the 'lag' output, depending on which of the lag/lead lines from the grating are activated, i.e. at logic '1'. This is shown in row F, where it should be noted that a tolerance of ±0·25 of the grating pitch can be tolerated on the shaft encoder signals.

This technique of lag/lead switch lines can be extended to the next decade to interrogate the next set of tracks on the encoder or to another encoder driven through a step-down gearbox. Fig. 4.17 shows the first stages of such a unit with a grating of 10 lines/mm of displacement, providing a resolution of 0·01 mm. Each set of shaft

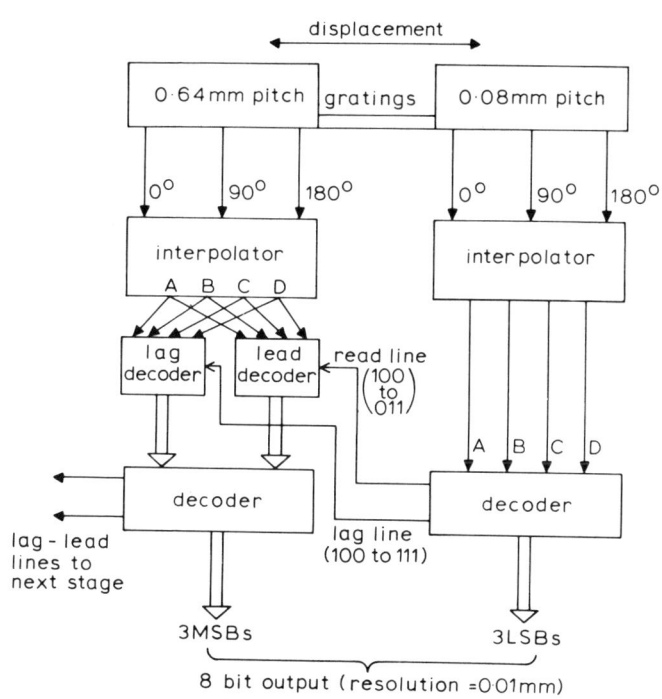

Fig. 4.18 Multitrack grating system

encoder tracks provides another decade. Thus, a total travel of 1 m, to a resolution of 0·01 requires five stages of shaft-encoder measurement but it would be providing a resolution of 1 part in 100 000 and only require mechanical assembly and tolerances which are easily obtained.

The method shown is for a decade readout, but, for purely digital readout, the interpolator output can be decoded to provide the required digital code and the shaft encoder output can be left in a digital code. The lag/lead signals are generated with respect to the main grating divisions which are the same as the least significant bits on the shaft encoder.

A system of this type, having a binary output has been constructed providing a 14 bit binary output, and a resolution of about 0·01 mm. It uses a grating with 5 lines/mm and an interpolator subdivision of 16. The standard commercial binary shaft encoder used provided 32 counts/rev and counted 32 revolutions. The encoder was driven by a lead screw having 1·6 threads per 10 mm.

The method has also been extended to the use of two gratings on the scale (i.e. a multitrack grating) one having a resolution eight times the other. The coarser grating replaces part of the rotary encoder. This configuration is shown in Fig. 4.18, in which it is seen that both gratings are provided with similar interpolators. The decoder for the three LSBs also provides the lead/lag lines for interrogation of the second and coarser grating output. This system provides a 6 bit binary output having a resolution of 0·01 mm for a grating pitch of 0·08 mm on the finer track.

4.4 References

NATIONAL ENGINEERING LABORATORY (1966): 'Absolute displacement transducer'
RUSSELL, A. (1966): 'An absolute digital measuring system using an optical grating and shaft encoder', *Instrum. Rev.*

Chapter 5

Analog conversion methods

This chapter deals with the problems encountered in instrumentation systems in which the output signal is required, finally, in digital form although the transducer output is in analog form. The subject matter is concerned with some of the aspects of equipment installation, the characteristics of multiplexers, filters, integrators etc. and their performance in sampled systems, and also with the characteristics of A/D converters.

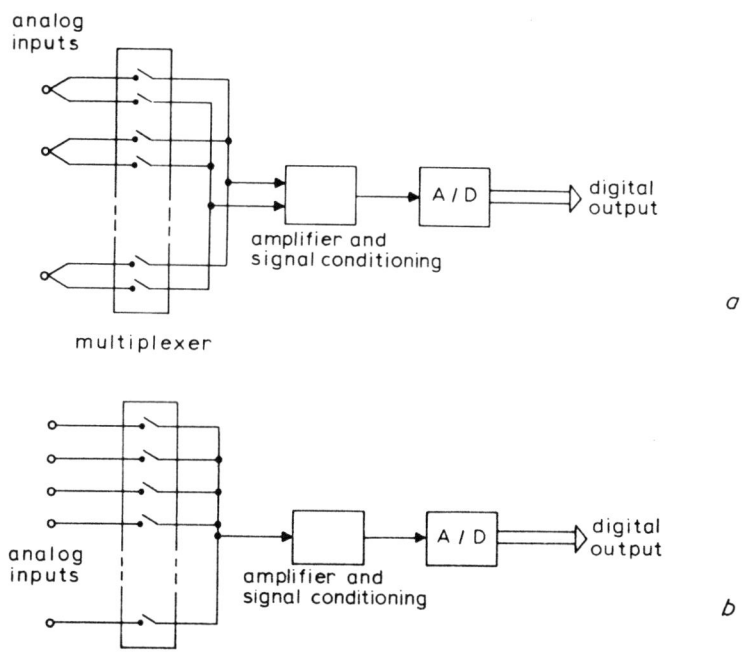

Fig. 5.1 Input circuits

5.1 Installation

Fig. 5.1 shows the general block diagram of a typical multiplexed analog conversion system in which a number of separate analog inputs are individually switched to a single A/D converter. One of the most significant problems arising in this type of installation are the unwanted signals which occur in most analog systems. These unwanted signals are caused by:

(a) noise, which may be any kind of electrical noise produced by power surges or switching transients in neighbouring equipment
(b) interference, which may be 'pickup' at mains frequency or its harmonics and, at times, from other sources
(c) crosstalk, which is the coupling of a signal from one source into the signal from another.

For convenience, these unwanted signals will all be classified as NIC (Sonnenfeldt, 1972) (noise, interference and crosstalk).

Although different in the manner in which they arise, NIC can be handled in a similar fashion. Every effort is made to minimise NIC at the source as it is more difficult to remove its effects on the signal than it is to prevent it arising. It is clear the NIC is a greater problem at small signal levels as in thermocouples, than in transducers developing large high-level signals. It is generally agreed that NIC can be reduced by careful selection of leads from the sources. Usually a twisted-pair in an insulated shielded sheath is used and, where possible, bundles of these shielded cables are installed in conduits for further physical and electrical protection. In this way electrostatic and electromagnetic pickup is minimised, and the insulation of individual cables prevents circulating current in the shield.

Special care must also be taken on the earthing of the separate cables, particularly by avoiding multiple grounds; generally shields should be earthed only at the transducer end.

Another contributing factor which must be considered in digital systems is that digitally-operated equipment produces its own NIC. This is due to the relays and solenoids in electrical typewriters, switch contacts, and digital output devices, e.g. stepper motors. It is therefore important in any overall system to ensure that self-induced NIC is minimised and isolated.

5.2 Multiplexers

Multiplexers belong to three main groups: rotary scanners, electromagnetic switches (reed switches or special purpose relays) and transistor systems. The primary difference between these types is the speed of operation, although other factors such as cost, introduction of NIC etc. are also important.. The selection of a particular type also depends on the type of A/D converter that is used. Generally, since a computer, data logger or a display can accept the data from only one transducer at a time, a single A/D converter is generally used, although consideration of the sampling rate required and the speed of A/D conversion may necessitate another configuration. If a high sampling rate is required, high-speed converters must be used, which will increase the overall cost.

In some circumstances, the effects of NIC can be minimised by averaging several samples from the same source. This can be done by arranging the multiplexer to read the same source for several samples and accumulating the sum in the computer or separate digital register where averaging can take place. An alternative arrangement is to use a special converter which has self-averaging output. In a very small system, i.e. systems with only a few input channels, it may be more economical to avoid multiplexing and use a separate A/D converter on each channel, selecting an appropriate converter according to the accuracy required and the magnitude of the input analog signal on each channel.

In many cases it is necessary to use two switches, one for each conductor of each analog signal, as shown in Fig. 5.1a. A third switch may be necessary in some installations to avoid 'earthing' problems and the introduction of NIC into the signal. For simplicity both these systems are usually drawn as shown in Fig. 5.1b.

Mechanical scanners

Rotary switches are cheap and often sufficiently reliable for simple installations. Typical of this type is the TS series of low-level scanners* (Pennington-Ridge, 1967). The scanner itself consists of a rotating arm carrying contacts which move over a switch disc containing 25 sets of gold plated contacts, each set comprising three radially displaced contacts, thus providing a 3-pole switching action. The drive motor is a thyristor controlled Slo-syn stepper motor which can be operated in either stepped or continuous modes at speeds that provide a sampling rate up to 10 channels/s. Switch noise and contact bounce are claimed to be negligible, being as low as $\pm 1\mu V$, hence enabling low-level signals (e.g. from thermocouples) to be switched. The maximum switching voltage can be as high as 100V.

*manufactured by IDM Electronics Ltd.

Channel identification is obtained from a coded contact disc driven by the switching shaft and is obtainable as a BCD output. This disc also provides the necessary information to enable the multiplexer to work in the addressable mode, in which the arm can be rotated to any position represented by a BCD address signal applied to the switch unit.

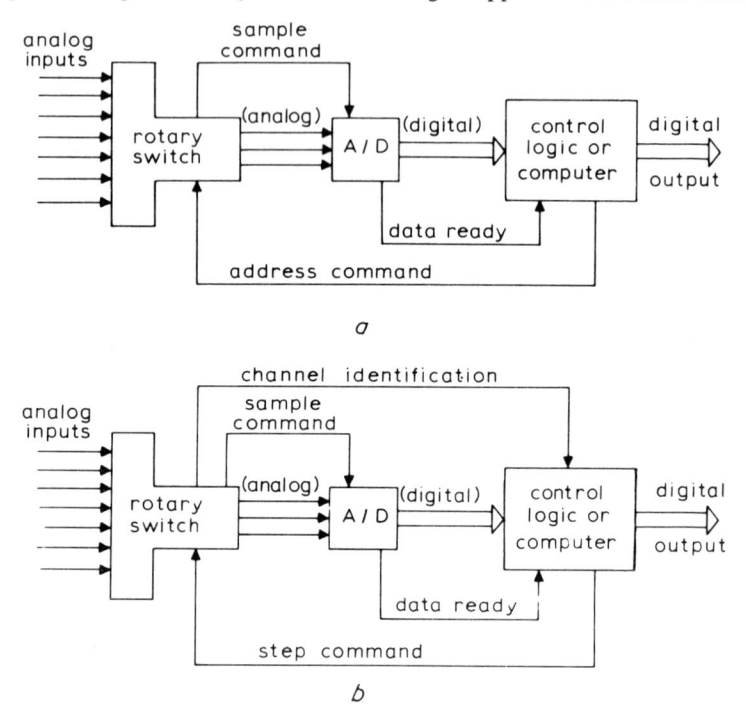

Fig. 5.2 Rotary mechanical multiplexer
   a   Addressable mode
   b   Continuous stepping mode

Working in conjunction with an A/D converter, the multiplexer operates as shown in Fig. 5.2. The upper diagram shows a system working in the addressable mode, the control logic or computer requesting data relating to a particular channel. The motor operates in the continuous mode, rotating the switch arm and the address disc, until the output of the disc is the same as the address from the computer. At this point the motor stops and a 'sample' command is passed to the A/D converter which makes the conversion to a digital signal and also passes a 'data ready' signal to the control logic or a computer.

In Fig. 5.2*b*, the scanner is shown working in the continuous stepping mode, the information passing to the control logic on both the channel identification information and the A/D output. The arm moves sequentially to the next channel after each A/D conversion and transfer of digital information has taken place. Alternatively, the control logic or computer could be programmed to increment the scanner only when required.

The only limitations of this type of multiplexing systems are in the speed of operation and in the usual dangers of contamination of switch contacts, causing increase in contact resistance, and also the generation of noise.

Multiplexers using relays

Some of the disadvantages of the rotary scanners can be overcome by the use of reed switches or relays. These usually consist of a pair of contacts in a sealed glass tube. The contacts are closed by energising a solenoid located around the tube. Conventional relays or, more specifically, mercury-wetted relays, are also used in a similar manner to reed switches. In either case, a small driver amplifier may be required to drive each solenoid, the appropriate amplifier being switched on by the control logic (Fig. 5.3a). More recently, reed switches housed in integrated circuit packages are available, designed to operate directly from TTL outputs and hence no special drive amplifiers are required. In this case, the control logic for switching can be derived from a logic decoder. The channel required appears as a binary of BCD-coded signal and the decoder provides an output, i.e. a logic '1', only to the chosen channel. Alternatively, the switches may be closed by clock-driven logic, which energises each solenoid in turn and also provides channel identification signals to the control logic. This system is shown in Fig. 5.3b and has a ring counter with a logic '1' on, say, the first channel output. The clock pulses progress this logic '1' through the ring counter successively switching on the other channels, one at a time. The logic output will also be used to develop a binary or BCD output to give channel identification.

For these systems, the A/D interface and its asssociated control logic will be similar to that described for the rotary scanner as similar command signals are required.

Reed switches do not suffer any contact contamination and are generally faster in operation than conventional relays and rotary scanners. Their low mechanical inertia and small electrical power requirements have permitted sampling speeds of up to 500 per second. Contact bounce and movement of the contacts in a magnetic field create noise, often lasting several milliseconds, following contact closure, and it is this which limits maximum sampling speeds.

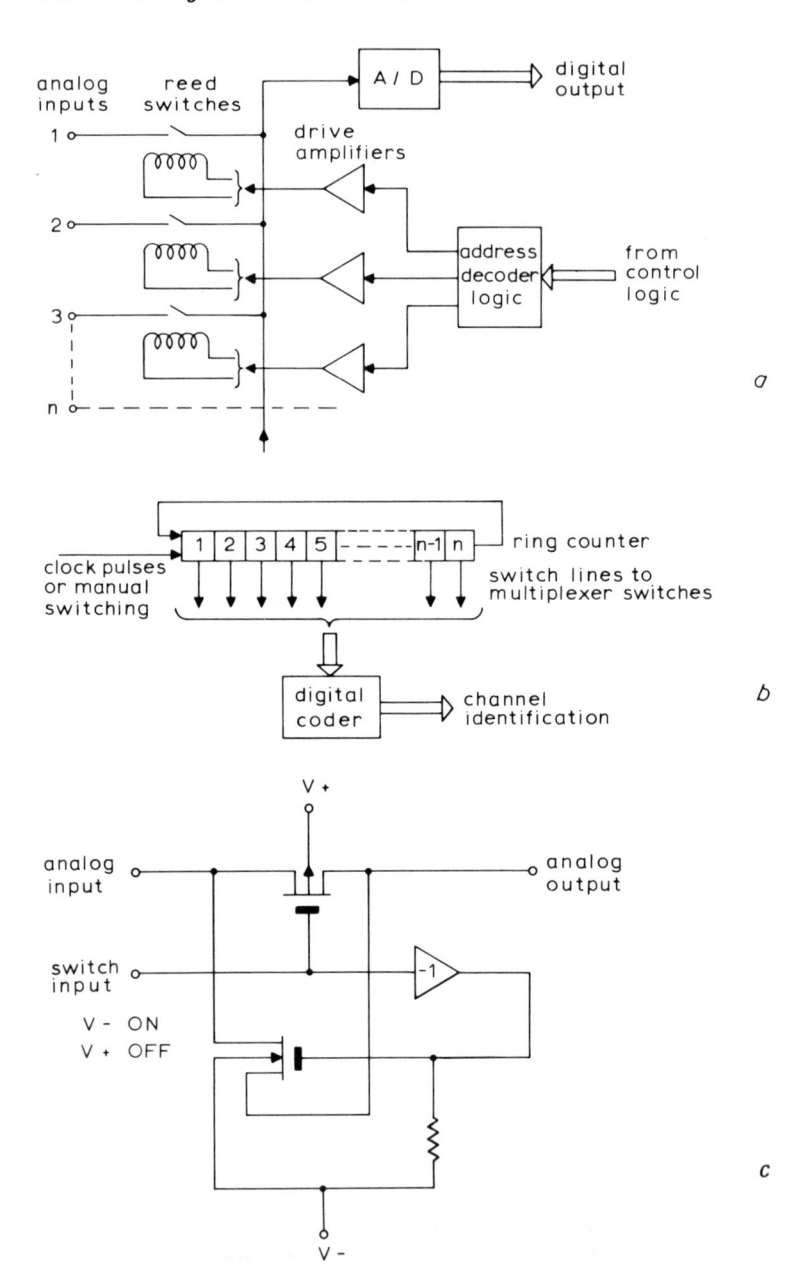

Fig. 5.3 Multiplexers
    a Reed switch multiplexer
    b Channel switching
    c CMOS multiplexer switch

Diaphragm relays can be used as an alternative to reed relays and are similar in that they have contacts which are sealed in a glass and metal capsule. One contact is fixed and located at the upper end of a metal core which protrudes from the glass seal into the energising coil. The other contact is located on a thin ferrous diaphragm, the edges of which connect to a ferrous seal ring forming part of the capsule. The diaphragm is insulated from the core by the glass seal but is connected (magnetically only) through the seal ring and a ferrous cover to the lower end of the core. When the coil is energised, the magnetic circuit is completed through the core, the ferrous cover and seal ring to the diaphragm, which is then attracted to the upper end of the core, causing the two contacts to close.

Commercial diaphragm relays are generally smaller than reed relays and have about ten times the contact force and much greater contact area. These and other factors tend to make the diaphragm relays less liable to contact bounce, more resistant to shock and vibration and have longer life than corresponding reed relays. They have, therefore, been used in multiplexing systems particularly associated with data logging systems.

Transistor multiplexers

Since it is essentially the mechanical parts of rotary scanner and reed switches that give rise to their drawbacks, it is obvious that some advantages can be obtained by using wholly transistorised units. This necessitates replacing each reed switch (and solenoid) by a transistor switch, the control logic remaining more or less unchanged.

Sampling rates can be up to many MHz although operating frequencies are often limited to about 100 MHz. Metal oxide semiconductor field effect transistors (MOSFETs) are becoming popular for use in multiplexers as they do not introduce offset voltages (Strassberg, 1972) and are more suitable for use in integrated circuit techniques than bipolar transistors.

Generally, transistorised multiplexers have a much longer life than those involving mechanical parts, although wetted-contact reed relays can have a long life, often sufficient to last the expected life of the complete installation. Transistor multiplexers can suffer from offset voltage drift primarily due to temperature effects. This does not usually occur in the other types of multiplexers.

A typical MOSFET multiplexer circuit is shown in Fig. 5.3c, only one channel being shown for convenience. When produced as integrated circuits containing a number of MOSFET switches the circuits are generally known as CMOS.

Noise from multiplexers

It is customary to show a multiplexer as consisting of one switch (or one pair of switches) for each channel closed, and switches for all other channels open. This is only true in the ideal multiplexer, in actual fact each switch has a series closed resistance and a finite open resistance as indicated in Fig. 5.4. In many cases the open resistance is very high, as in rotary scanners and reed switches, often but not quite so high in transistor switches. Similarly, closed resistance is generally lowest in rotary scanners and reed switches, and highest in some transistor switches.

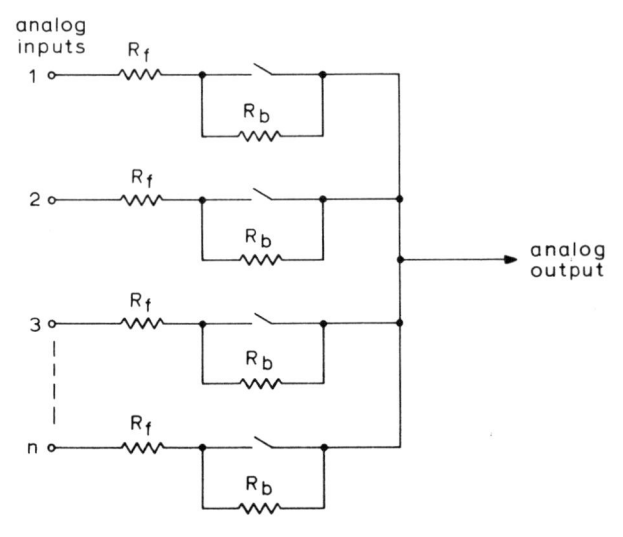

Fig. 5.4 Multiplexer switch resistance
R_f = closed resistance of switch
R_b = open resistance of switch

The open resistances together with the transducer impedances of all open-switches are in parallel with the signal path through the one closed switch with its low resistance. When the number of channels is large, the resultant 'open resistance' path could be comparable to the resistance of the single closed switch. This can often result in a high degree of crosstalk and it also acts as an additional path for all NIC generated in all the connections to the multiplexer. This effect can only be minimised by using separate banks of multiplexers with a limited number of inputs on each, which has the effect of reducing the total number of 'open resistance' paths in parallel.

5.3 Signal conditioning

Signal conditioning is necessary to suppress noise etc. while transmitting sufficient information relating to the signal. Signals which are processed through a multiplexer are essentially samples and therefore appear as pulses of amplitude A and duration a (Fig. 5.5). The duration a is the sampling time, that is, the time that the appropriate switch is closed by the multiplexer. If the signal is varying, the amplitude could be varying during this sampling time unless a sample and hold circuit is used before the multiplexer. By using the Fourier transform pair (Duckenfield, 1965):

$$F(\omega) = \int_{-\infty}^{+\infty} f(t)e^{-j\omega t} dt \qquad (5.1)$$

$$f(t) = \frac{1}{2\pi} \int_{-\infty}^{+\infty} F(\omega)e^{j\omega t} d\omega \qquad (5.2)$$

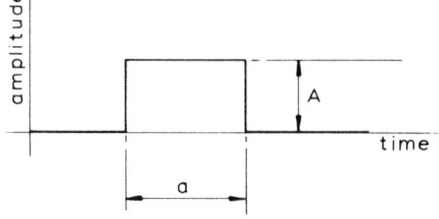

Fig. 5.5 Sampled pulse

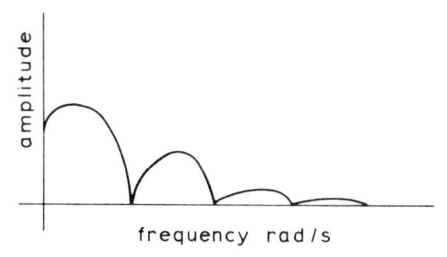

Fig. 5.6 Frequency spectrum

the pulse of Fig. 5.5 can be represented by the frequency spectrum shown in Fig. 5.6. It is seen that most of the spectral energy of the pulse occurs in the low-frequency region which suggests that a suitable filter for suppressing NIC would be a 'low-pass' filter which would attenuate the high frequencies (that also arise in NIC) without seriously reducing the information in the signal pulse.

Low-pass filter

Certain difficulties arise in selecting a suitable low-pass filter and, as in most engineering design problems, a compromise has to be selected from conflicting requirements. Some of the problems can best be explained by considering the simple passive low-pass filter shown in Fig. 5.7a. In practice active filters would in fact be preferred.

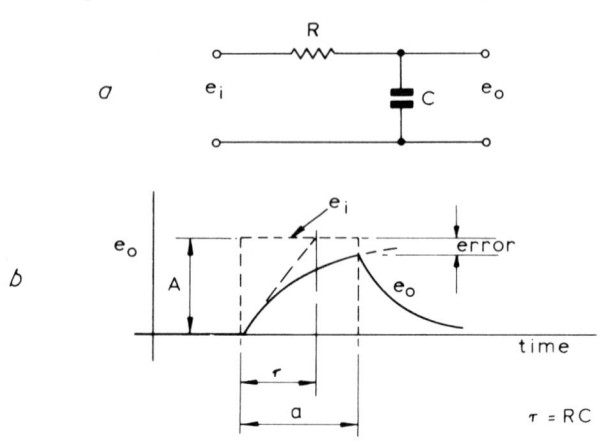

Fig. 5.7 Lowpass filter and response
    a Lowpass filter
    b Filter response to a pulse

If the input e_i is a pulse of amplitude A, then

$$e_0 = e_i(1 - e^{-t/\tau}) \quad \text{where } \tau = RC \tag{5.3}$$

This has the waveform shown in Fig. 5.7b and hence the output will not reach the required amplitude A in the sampling time a unless a suitably fast time constant τ is selected. The error in the system at the end of the sampling time a is given by

$$\text{error} = e^{-a/\tau} \tag{5.4}$$

Thus the measurement error can be related to time constant in the following manner

% error	a/τ
10	2·2
5	3·0
2	3·9
1	4·6
0·5	5·3
0·1	6·9
0·05	7·6
0·01	9·2

This means that if the error is not to exceed 0·05%, then the sampling time a must be 7·6 times the time constant τ of the filter.

Consider now the attenuation properties of the filter. Although noise contains a high proportion of high-frequency components, interference is often caused by the electrical mains and the induced voltages which occur are at line frequencies, i.e. at 50/60 Hz, and special attention should therefore be given to high attenuation of these frequencies. The amplitude ratio of the simple filter at any frequency is given by:

$$\left(\frac{e_0}{e_i}\right)_\omega = \left|\frac{1}{1+j\omega\tau}\right| = \frac{1}{\sqrt{(1+\omega^2\tau^2)}} \tag{5.5}$$

If $\omega = 50$ Hz $= 314$ rads^{-1} and it is required to have an attenuation of 10 : 1 (that is NIC at 50 Hz, and all frequencies greater than 50 Hz are attenuated by the filter to at least 1/10 of the magnitude at the input) then:

$$\left(\frac{e_0}{e_i}\right)_{50} = \frac{1}{\sqrt{(1+314^2\tau^2)}} = 0\cdot 1$$

Hence $\tau = 0\cdot 031$s.

This is the minimum τ which must exist to achieve the required attenuation and this is often known as a waiting characteristic.

In a filter which has a requirement of a measurement accuracy of 0·05%, the sampling time a must be at least 7·6 times τ. Therefore the sampling time must be 0·236 s (i.e. 7·6 x 0· 031) if both accuracy and attenuation requirements are to be met. This indicates that the sampling rate should not exceed four per second which would be too slow for most practical systems.

Elaborate filters can be designed to give more efficient attenuation but there is always a minimum time before a given tolerance (i.e. minimum error) in the measurement can be achieved. Hence, the basic argument of the simple low-pass filter enumerated above, still applies and alternative methods of minimising 50/60 Hz interference must often be adopted.

It can be shown that the response of a low-pass filter of the simple type shown in Fig. 5.7a to a step input, is of the form:

$$e_0 = \frac{1}{\tau}\int e_i dt - \frac{1}{\tau^2}\iint e_i dt + \frac{1}{\tau^3}\iiint e_i dt - \ldots \tag{5.6}$$

and by ignoring the higher-order term this can be approximated to

$$e_0 = \frac{1}{\tau}\int e_i dt \tag{5.7}$$

This has given rise to the term of 'integrator' being applied to these filters and the terms 'low-pass' and 'integrator' are often interchangeable in this context.

Position of filters

Filters can be placed before or after the sampling multiplexer, and the required characteristics of the filter are somewhat different for these two cases. Consider a system with S inputs and switches, each with a sampling period of a seconds; then, assuming sequential switching, the period between successive samples from each individual input is at least aS seconds, as a finite time is necessary to switch off one channel before the next one is switched on. If the sampling rate is to be sufficiently frequent to record all significant changes from each sampled input, then, from the sampling theorem, the sampling rate ($1/aS$) must be at least twice the highest significant frequency component f_h contained in the output.

That is: $\quad \dfrac{1}{aS} > 2f_h \qquad (f_h \text{ in Hz})$

$\quad$ or $\quad \dfrac{\pi}{aS} > \omega_h \qquad (\omega_h \text{ in rads/s}) \qquad (5.8)$

If the filter is placed ahead of the sampling switch it need pass only frequencies up to f_h for that particular channel.

If the filter is placed after the multiplexer then it must be selected so that a sufficient number of filter time constants elapse during the sampling interval to achieve whatever accuracy is required. Thus, the pass-band requirements are considerably different.

As an example, consider the previous case where an accuracy of 0·05% was required, the time constant of the filter τ is given by $\tau = a/7·6$ for a simple RC filter *following* the sampling switch. The passband of the required filter (ω_{pi}) can be defined by:

$$\omega_{p_1} = \dfrac{1}{\tau} = \dfrac{7·6}{a} \text{ rads/s}$$

If a filter is placed *before* the sampling switch then the required passband ω_{p_2} must be at least equal to ω_h, where ω_h = frequency of highest component in signal.

Hence $\quad \omega_{p_2} = \omega_h = \dfrac{\pi}{aS}$

$$\frac{\omega_{p_1}}{\omega_{p_2}} = \frac{7 \cdot 6}{a} \frac{aS}{\pi} = 2 \cdot 4S \qquad (5.9)$$

Thus the passband of the filter after the sampling switch must be $2 \cdot 4S$ times as large as that of a filter situated before the sampling switch. As the noise attenuation of these filters depends on the passband, i.e. the smaller the passband, the greater the noise rejection (in general), then filters placed before the sampling switch have considerable advantage, particularly for systems where S may be large (greater than 100). However, the cost increases proportionally to the number of filters and this must be considered in the overall system design.

For small systems, up to, say, 10 inputs, pre-sampling filters are preferred because of their noise-rejection advantages. For medium sized systems, two or more stages of sampling may be adopted as a compromise between the many conflicting requirements.

Amplifiers

Amplification of some analog transducer signals may be necessary to allow the use of a single range A/D converter and to retain the maximum possible accuracy. Some converters can be programmed to automatically change range according to the signal level, but this requires more programming complications if the system is controlled from a computer.

In many cases, the amplifiers are located on the signal lines to the multiplexer, particularly if the number of channels is small. Alternatively, a single amplifier may be placed after the multiplexer, but the impedance characteristics of the multiplexer must be considered as part of the input circuit to amplifier.

Signal filtering

The following notes refer to amplifiers and any other apparatus to which the signal source is connected. This could be the multiplexer input, or, in unmultiplexed systems, the A/D converters. Many factors cause NIC, and, even if the transmission link is not susceptible to 'pick-up', the transducer themselves can be the cause of unwanted noise. The most critical conditions generally arise in thermocouples where the different wires which are necessary, cause unbalanced pick-up that becomes superimposed on the signal input to the amplifier.

Noise generated in the conducting wires of the transducer is often termed normal (or series) mode interference.

The other major source of pick-up is usually common earthing points and is called common-mode noise. It also arises in any circuit

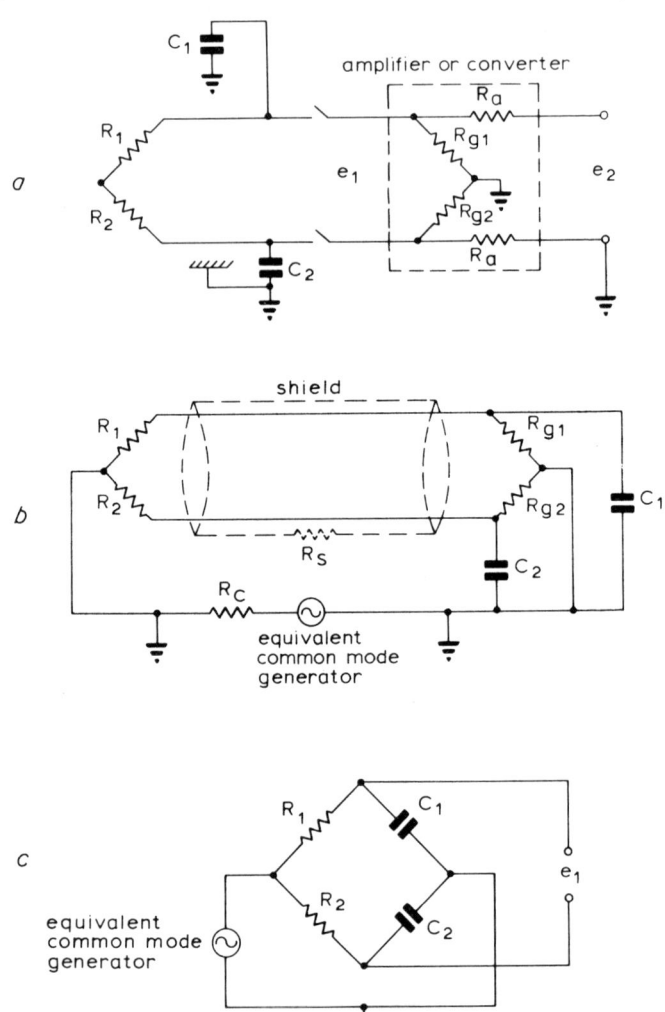

Fig. 5.8 Input circuits
    a   Thermocouple installation
    b   Grounded source
    c   Equivalent circuit for conversion of common mode to normal mode

that is not completely balanced (resistive and capacitive) with respect to earth at every point in its length. Screened lead (guarding) can minimise these effects. A typical low-level signal as from a thermocouple, together with the associated stray capacitances and amplifier input impedances is shown in Fig. 5.8a.

The induced voltages are considered as 'a common mode equivalent generator' as shown in Fig. 5.8b and it has been shown that this may reach a value of up to 10V in bad circumstances. The resistances Rc and Rs are normally very low compared to other resistances in the equivalent network and have little influence. Usually, the amplifier has differential inputs with impedances Rg_1 and Rg_2 as shown.

At very low frequencies, the stray capacitances C_1 and C_2 between the leads and earth can be neglected and the common mode NIC developed along the leads (resistances R_1 and R_2) will be considerably attenuated at the input terminals, providing the input impedances Rg_1 and Rg_2 are large. With suitable transistor inputs, input impedances are in the order of 10^8 Ω, compared to values of 10^2 Ω, for R_1 and R_2. Thus, the common mode rejection of $10^6:1$ (120 dB) is obtained at very low frequencies. It must be remembered that using a differential amplifier, only unbalanced voltages in R_1 and R_2 will cause NIC at the amplifier input.

For a.c. voltages the stray capacitances C_1 and C_2 become important, and induced voltages (NIC) will appear at the input of the amplifier as 'normal mode' signals since the capacitances reduce the effective input impedances.

It is easily shown with very low signal levels, as developed by thermocouples, that the common-mode voltage can exceed the signal voltage, and filtering the signal before it reaches the amplifier, or providing filtering in the amplifier itself, is vital.

Many of these problems do not arise where the transducers produce larger signal levels than normally associated with the thermocouples. Signal levels measured in volts rather than millivolts are easier to handle, and NIC represents smaller possible errors.

A further difficulty arising in amplifiers is the recovery time after overload. Overload can often occur due to harmonics of the mains frequency (50 Hz) which see a lower impedance and therefore appear as high voltages at the amplifier input. Another cause of overload occurs when a multiplexer fails to close both lines of the signal source at the same instance. In this case all of the NIC is applied to one side of the amplifier while the other side is temporarily floating. This is compared to more usual cases where NIC results from the difference in line resistances ($R_1 - R_2$) only. A fast recovery time from these overloads is essential if the sampling speed is to be maintained.

Digital filtering

One of the inherent advantages of digital and computer controlled systems is that the computer can be used to control the sampling and also the manner in which the resultant values are manipulated. Digital filtering is one method of control whereby some of the effects of NIC voltages, which consists mainly of mains frequencies (50 Hz) and its harmonics, can be minimised.

Assume that the desired signal is essentially a d.c. voltage e_s on which the NIC is superimposed. The continuous signal can be represented by

$$e_0 = e_s + e_1 \sin \omega t + e_2 \sin 2\omega t + e_3 \sin 3\omega t + \ldots \quad (5.10)$$

If a sample is taken at t_1, the digital voltage e_{d_1} accepted by the computer will be the digital equivalent of

$$e_{d_1} = e_s + e_1 \sin \omega t_1 + e_2 \sin 2\omega t_1 + e_3 \sin 3\omega t_1 + \ldots \quad (5.11)$$

If a second sample is taken at t_2 such that

$$t_2 = t_1 + \frac{\pi}{\omega}$$

then the digital voltage will be

$$e_{d_2} = e_s + e_1 \sin(\omega t_1 + \pi) + e_2 \sin(2\omega t_1 + 2\pi) + \ldots \quad (5.12)$$

Both these samples can be stored and averaged by the computer (i.e. added and divided by 2) giving rise to a resultant digital voltage:

$$e_{d_v} = \frac{e_{d_1} + e_{d_2}}{2} = e_s + e_2 \sin 2\omega t_1 + \ldots \quad (5.13)$$

In this way the resultant consists of the genuine signal voltage plus the even harmonics (2ω, 4ω etc.) of the NIC; the fundamental and all the odd harmonics being suppressed completely. As in practice most of the NIC signal is at the fundamental frequency of the power source, this method has considerable advantage. Further, if filtering of the signal source is still incorporated, it is now only necessary to filter frequencies of 2ω and greater, which, as explained earlier, is simpler and allows higher sampling rates than if filtering of the fundamental is also required.

This method can be extended into a more thorough averaging procedure, sometimes referred to as integration. If the signal e_0 is sampled regularly over a period T, the average value can be computed from:

$$e_{av} = \frac{1}{T} \int_0^T e_0 dt \qquad (5.14)$$

If e_0 has the same form as eqn. 5.10 then

$$e_{av} = e_s + \frac{1}{T} \left\{ \frac{e_1}{\omega}(\cos \omega t - 1) + \frac{e_2}{2\omega}(\cos 2\omega - 1) \right. \\ \left. + \frac{e_3}{3\omega}(\cos 3\omega t - 1) + \ldots \right\} \qquad (5.15)$$

By making T sufficiently long, the NIC terms in the curly brackets can be made relatively small. However, the integration interval T cannot exceed the sampling period, unless some programming allows a separate integration for each channel within the computer.

In some special purpose A/D converters, provision is made to perform this 'integration' by making many conversions and averaging during the time that the signal is connected to the converter, i.e. during the sampling period. A typical converter of this type can provide an almost infinite rejection of 50 Hz and 100 Hz NIC for selected sampling speeds. However, they suffer the basic limitation of this method, which is, the shorter the sampling period the less effective (i.e. more selective) does the NIC rejection become.

5.4 Analog-to-digital converters

Providing an analog voltage of a parameter to be measured can be obtained, a digital equivalent may be obtained by one of the standard A/D techniques. However, the technique selected for a particular purpose depends on many factors such as speed of conversion, accuracy, cost etc. It is assumed that the analog signal is continuously available, that it is connected to the input terminals for a sufficient time for an A/D conversion to be made and is at least to the required degree of accuracy of the final digital representation.

A/D converters are themselves sampling devices since a given digital output represents an instantaneous value of the analog signal. The converter will also take a finite time to make the conversion and the digital output is therefore usually a delayed representation of the analog signal. The time available for conversion is usually less than the duration of each sample due to unwanted transients at the beginning of the sample arising from the multiplexing action. This does not necessarily arise if the converter is permanently connected to a single analog signal.

Aperture time

A most important characteristic of any A/D converter is the aperture time, which is the time interval during which the converter is sensitive to the input signal. The aperture time required depends on the acceptable error arising from an input signal changing in magnitude during the digitising process. In practice it is taken as the time during which a worse case signal changes by one least-significant-bit (LSB).

Consider a signal of frequency f Hz which is converted to an N bit binary digit number and which changes by one LSB as the input signal passes through zero. The aperture time T_A is given by

$$T_A = \frac{1}{\pi f 2^N} \tag{5.16}$$

The digitising time T_D is the time required by the converter to provide a digital output when the input is at maximum. Aperture time is usually less than the digitising time.

Accuracy of conversion

The accuracy of conversion (i.e. the greatest degree of accuracy obtainable) from analog to digital is limited by the 'quantising error' which depends on the number of binary digits and the maximum rate at which these can change. Quantising error is given by

$$e_q = \pm 0\cdot 5 \; \frac{1 + kT}{2^{N-1}} \tag{5.17}$$

Where T_A represents the aperture time and is the total time required for a change of one digit, k is the rate of change of analog input signal in LSB/s.

The foregoing equation indicates, as might be expected, that quantising error varies linearly from 0·5 LSB to 1 LSB as the rate of change of input k increases from zero to 1 LSB per aperture time interval. At input rates above this, the law holds for some conversion techniques while, in others, the rate increase of error rises.

Sources of error

The actual accuracy of conversion is usually worse than that defined by the quantising error of eqn. 5.17 due to a number of factors. These are:

(a) noise error
(b) nonlinearity error
(c) gain error
(d) zero or offset error

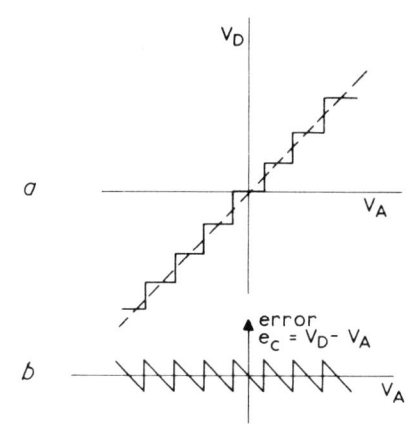

Fig. 5.9 Ideal conversion

In an ideal conversion, the relationship between the actual analog equivalent of the digital output V_D and the real value of the input to the converter V_A can be represented by the diagram shown in Fig. 5.9a. Since the conversion error is given by:

$$e_c = V_D - V_A \qquad (5.18)$$

and e_c varies by ±0·5 LSB, it can be represented by the diagram shown in Fig. 5.9b. Each of the errors listed above can be examined by reference to the ideal case given in Fig. 5.9.

Noise error

The error due to noise is that which appears as an output due to the noise generated in the converter only and not that present in the input signal. However, it is convenient to refer this noise error to the input and it may then be considered as a zero shift, i.e. V_D will be raised (or lowered) vertically with respect to the ideal, and this is shown in Fig. 5.10a with the corresponding error shown in Fig. 5.10b.

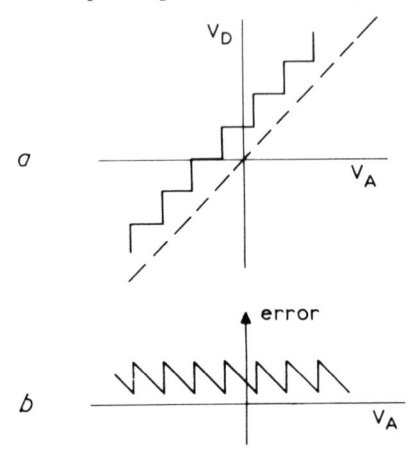

Fig. 5.10 Zero error or noise error

Similar in effect to noise, is the error due to recovery time after overloading the converter. In converters where the comparator recovery is exponential from an overloaded condition, then, if a conversion is made too soon, a zero error results. Since this is spread over a limited range, it can be considered as a form of noise.

Nonlinearity error

Nonlinearity in converting V_A to V_D occurs if the converter adds (or subtracts) an extra digit for a given change in V_A. An example of nonlinearity is the hysteresis effect shown in Fig. 5.11a and the corresponding error in Fig. 5.11b. A second form is shown in Fig. 5.11c and Fig. 5.11d, which can result in the digital number oscillating between two values.

Gain error

The converter gain can change due to deterioration of the scaling resis-

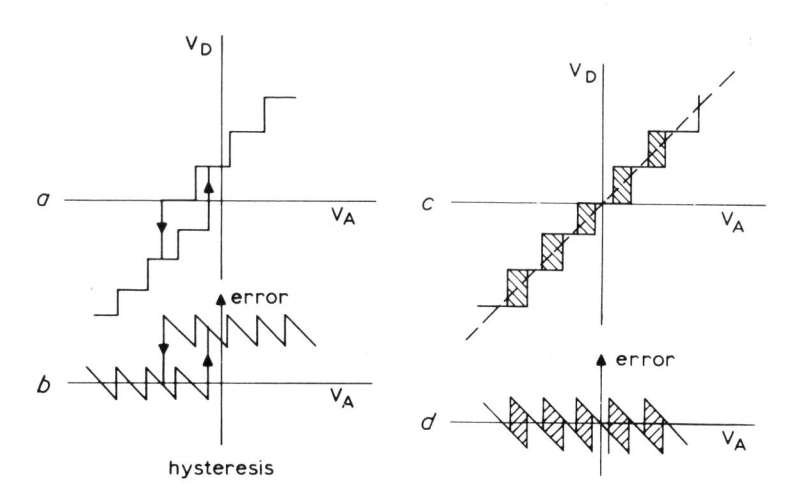

Fig. 5.11 Nonlinearity errors

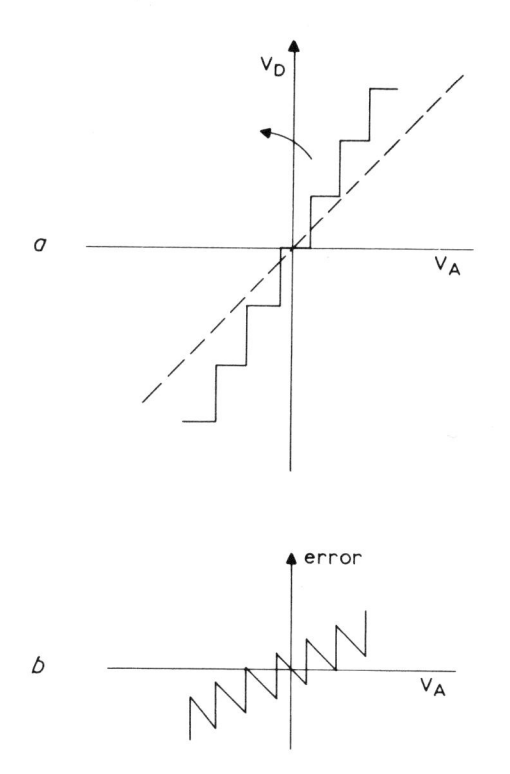

Fig. 5.12 Gain error

tors or, more possibly, by alteration of the reference voltage. Each of these will alter the $V_A - V_D$ slope and hence give an error proportional to V_A. This is shown in Fig. 5.12a and the corresponding error in Fig. 5.12b.

Offset error

Offset errors are usually referred to the input, as in noise error. If an output voltage V_D is present for a zero input V_A, and this remains a constant value over the entire range, it can be classified as an offset error and can be represented in the same manner as a zero shift, as shown in Fig. 5.10a and 5.10b.

Conversion techniques

The selection of a particular conversion technique will depend primarily on the requirements of accuracy, aperture and digitising times and cost, but power-supply requirements and environmental conditions may also be important considerations in many cases. There are two basic techniques now used for converters. These are:

(a) analog methods
(b) feedback methods

These will be discussed in relation to the specific requirements mentioned above.

Analog methods

The most commonly used analog method is the one in which the analog input voltage is first converted to a pulse of time, length proportional to the value of the input voltage. The converter requires an accurate clock which produces a regular stream of pulses and the digital output is in effect the number of clock pulses that are gated by the length of the pulse developed from the analog input.

A block diagram of the method is shown in Fig. 5.13. The voltage pulse of duration T_p is generated by comparing the input voltage V_A with a linear ramp voltage that can be generated by an operational amplifier connected as an integrator. The constant-frequency clock pulses are gated into a counter-register for the duration T_p of the pulse. The count in the register continues until the ramp voltage equals V_A, thus generating the end of the pulse. The final count at the end of the pulse represents the analog input in digital form. The maximum digitising time is $2^N \tau$, where N is the number of bits required and τ the clock period. The aperture time is also $2^N \tau$ for complete digitisation.

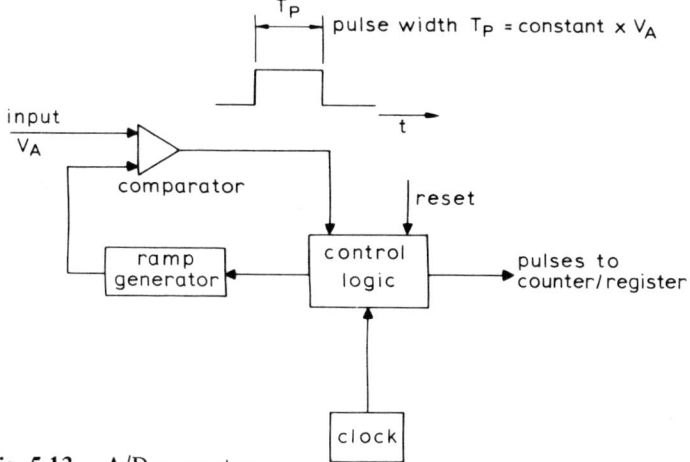

Fig. 5.13 A/D converter

Relative to the clock period these times are long, and hence limit the application of this technique. Improvements are possible which reduce the aperture time to more acceptable values, but other techniques of A/D conversion are often preferable.

Analog systems therefore have a slow conversion rate and have medium accuracy (generally $> 0 \cdot 5\%$) but they do have the advantage that they are simple and therefore cheap. This can be used to advantage in computer-controlled systems where only few inputs are required

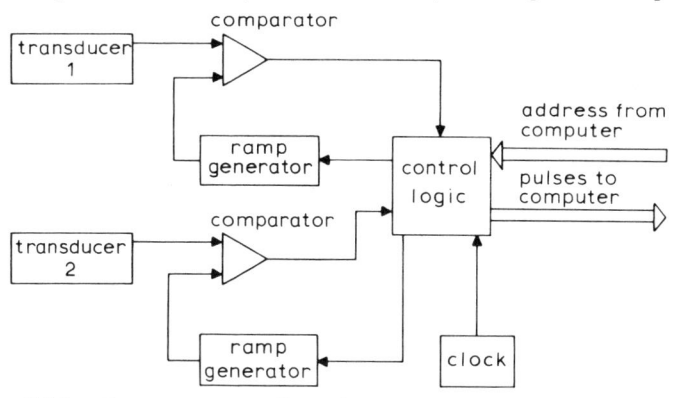

Fig. 5.14 Computer-controlled A/D conversion

and great accuracy is not demanded. A possible scheme is outlined in Fig. 5.14 for two separate transducer inputs, each transducer having its own ramp generator and comparator but requiring only a single clock and register, although the register could also be incorporated in the computer.

Address logic will decode the information from the computer and pass a fixed-level signal to the appropriate ramp generator on the selected transducer. At the same time the clock pulses will start to feed into the register or counter. When the ramp voltage equals the transducer analog signal, the comparator output changes, sending a high-level signal to the control logic, stopping the clock count in the register or computer. This system gives a cheap method of effective multiplexing and A/D conversion with only fixed high-level signals transmitted to and from the transducer point to the computer, thus avoiding many of the NIC problems discussed earlier.

The ramp necessary for this method can be replaced by a staircase waveform which increases the overall accuracy. This is because the accuracy is no longer dependent solely on the clock but on the components used to generate the staircase. A further variation is the voltage/frequency method in which the frequency of the output is proportional to the analog input V_A, (see also Chapter 3). The frequency then is measured by counting the number of cycles during a given time interval. This could have application to interrogative transducers used in online systems, particularly if part of the frequency generating circuit was actually attached to the transducer.

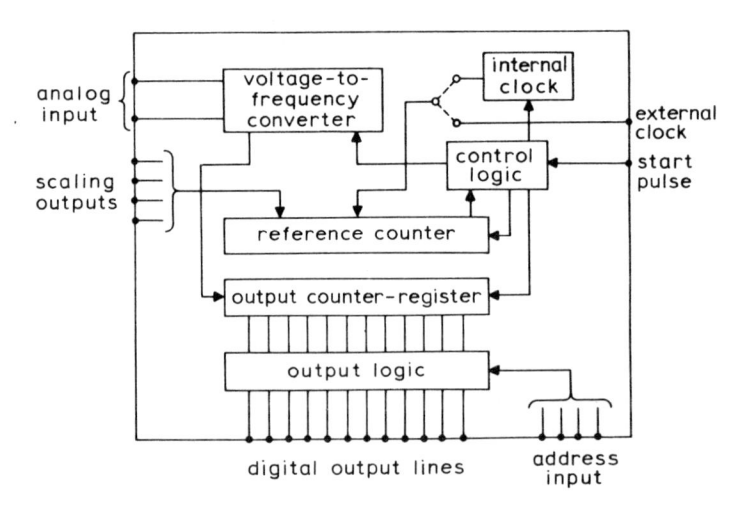

Fig. 5.15 A/D converter (voltage/frequency)

A/D converters of this type have been produced by a number of manufacturers, especially for use in computer controlled systems. The block diagram of one particular unit is shown in Fig. 5.15. The analog input voltage is connected directly to a voltage/frequency converter (VFC) of the type discussed in Chapter 3. The output is a train of pulses whose frequency is proportional to the input voltage and of duration equal to the clock frequency. In the panel shown, the clock may in fact be some other timer outside the converter itself.

A measurement is started by a start pulse from the computer which causes pulses from the VFC to accumulate in the reference counter directly from the clock. Simultaneously, pulses from the VFC are counted into the output counter register. The reference counter is used for scaling purposes and when this counter reaches the level determined by the scale setting, indicating a given lapse of time from the start pulse, the count into both registers is stopped. The binary number in the output register then represents the input analog voltage to a specified scale and the register may be switched into the data lines. A further refinement is that the output is transferred to the output lines only on receipt of the appropriate address input logic which represents that particular converter. Both registers must be cleared before a new conversion is required.

Typical performance of this type of unit will give an overall accuracy of about ±0·1% and have an input impedance of 4MΩ. Generally, the overall accuracy depends on the VFC and the clock; with some units the accuracy can be improved to ±0·05% over a temperature range of 0°C to 55°C. Using MOSFET integrated circuits the input impedance can also be raised to 100 MΩ or more.

Feedback methods

All feedback methods of A/D conversion rely on a D/A converter to develop an analog voltage proportional to the digital output. This is the feedback which is compared with the input voltage and the difference used to control the A/D conversion.

Typical of the D/A converter used in these systems is that shown in Fig. 5.16a in which an analog summing amplifier is used to sum suitably weighted proportions of a reference voltage V_R. This reference voltage must remain constant at a known value and this is achieved by using standard cells or zener diodes. The output of the amplifier is the sum of the voltages represented by all those input resistors where the switches are closed. For example, if the switch S_{N-1} is closed then

$$V_0 = \frac{R}{R} V_R = V_R$$

If switch S_{N-3} is closed, then $V_0 = \frac{R}{4R} V_R = 0\cdot 25\, V_R$

If both S_{N-1} and S_{N-3} are closed, then $V_0 = 1\cdot 25\, V_R$

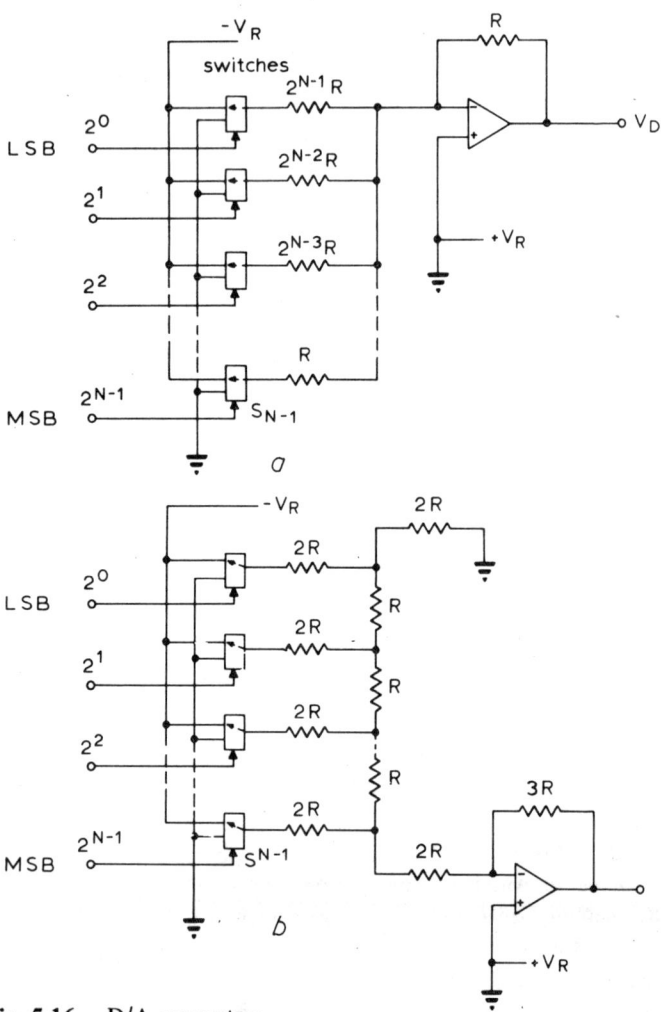

Fig. 5.16 D/A converters

In this way the digital input signal turns on those switches on which the logic input is '1' and the resultant output voltage is the sum of the appropriate voltages and therefore a direct equivalent of the digital input, providing the resistors have the ratios shown for a straight binary input. The overall accuracy of the system depends on the stability and accuracy of the reference voltage and the resistors.

An alternative method is the ladder-type converter shown in Fig. 5.16b which operates in a similar fashion but now has only a limited range of resistors which reduces some of the stability problems that can arise with high value resistors.

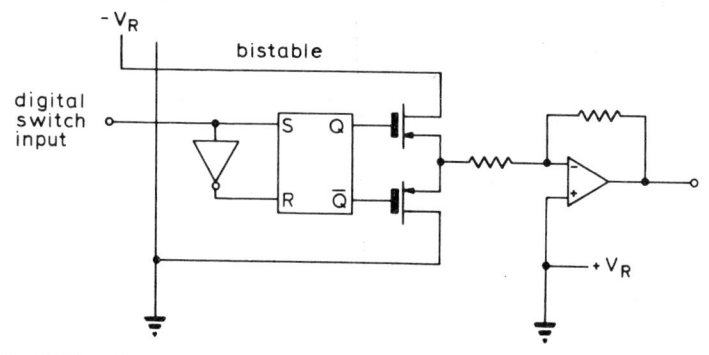

Fig. 5.17 Digital switch

The switching used in these converters usually incorporates a bistable and MOSFETs, and a typical arrangement is shown in Fig. 5.17, which shows just one switching station. Normally, the bistable and MOSFETs are produced as complete integrated circuits. In fact integrated circuits are available that contain all the components for a D/A converter except the resistors, which are connected externally.

Returning to the overall feedback A/D converter, Fig. 5.18 shows the basic feedback arrangement in which the D/A is used to develop an analog signal proportional to the digital output. This signal is compared to the input analog voltage V_A to be converted. The output

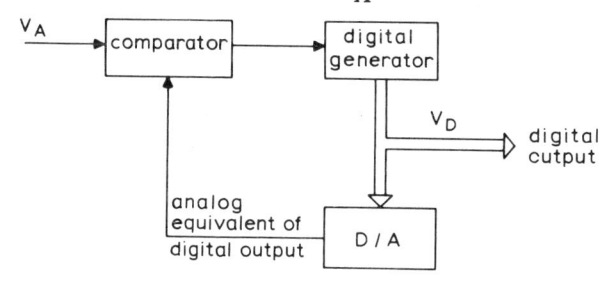

Fig. 5.18 Feedback A/D

of the comparator is used to control the formation of the digital output. When the analog signal developed by the D/A converter is equal to V_A, the comparator output will hold the digital output at a fixed value which represents the digital equivalent of the input voltage.

There are a variety of ways in which the comparator output is used to develop the correct digital output. Usually, each involves a clock to generate pulses, a counter and a control unit, although they may not always be identified as separate units. The method chosen usually decides the digitising and aperture times. Only three of the common methods will be described here:

(a) digital ramp
(b) reversible counter
(c) successive approximation

Digital ramp

This is a modified and improved form of the ramp method described earlier and shown in Fig. 5.19. The comparator output is used to start the count of clock pulses into a counter until the digital output is equivalent to input voltage V_A. At this instant the comparator voltage is zero and the count will stop. It is necessary for the control logic to clear the counter to zero before each measurement. The digitising time T_D has a maximum value of $2^N \tau$ as each measurement will start from zero digital output. Similarly, since only an increase in the count can be accepted, the aperture time is also $2^N \tau$.

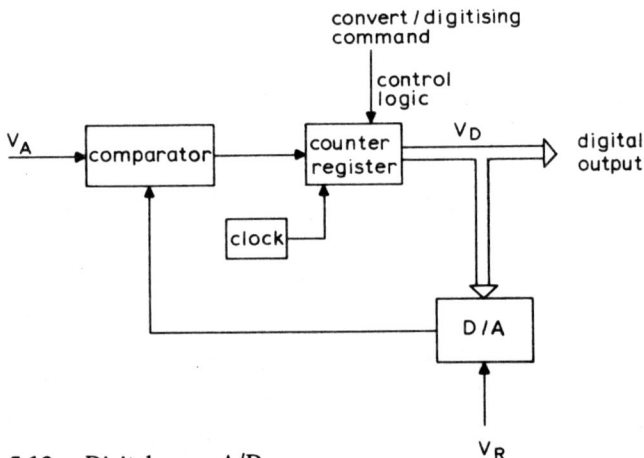

Fig. 5.19 Digital-ramp A/D

Reversible counter

This technique overcomes the long aperture time by allowing the counter to count up and down according to the output of the comparator indicating the digital output is less or greater than the input voltage to be measured. In this way, the digital output will follow the input continuously, and the aperture time is reduced to τ, i.e. the clock rate. If a convert pulse is used to zero the counter, then the maximum digitising time will still be $2^N \tau$.

The control logic can be arranged to allow the counter to continuously following the input voltage and a 'read' pulse supplied to transfer the digital number in the counter to a separate output register as required.

If the input signal is varying by a rate higher than $1\ \text{LSB}/\tau$, then the counter cannot follow and the quantising error will be greater than that given by eqn. 5.17.

Successive approximation

The block diagram of this system is the same as for the reversible counter, i.e. Fig. 5.19, except that a shift register is used in the control logic. This is the most widely used technique at the moment since it has fast digitising and aperture times.

A digitising command resets the digit output to zero and internal logic then develops a logic '1' on the line representing the MSB. The comparator compares the input voltage V_A with the analog voltage

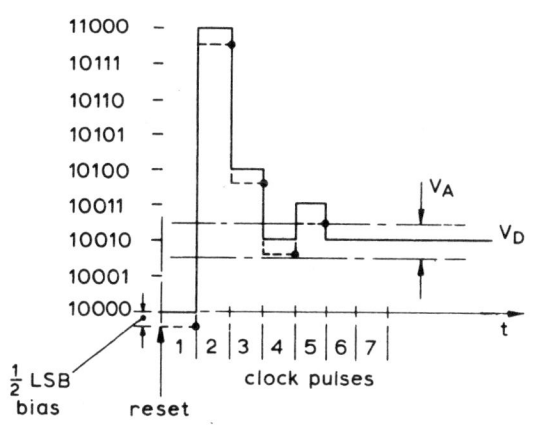

Fig. 5.20 Successive approximation

developed by the D/A when the MSB switch is closed. If the comparator output is positive, indicating that $V_D > V_A$, the control logic removes logic '1' from the MSB line and tries logic '1' on the next MSB. The process continues successively at a rate dictated by the clock, logic '1' remaining on any line whenever $V_A > V_D$. This process is illustrated in Fig. 5.20. Quantising error for varying signals is given by eqn. 5.17 for all input rates up to the point where the switching transients in the logic circuits start to affect the performance.

Conclusion

In all the above methods, and in many others, the arrangement of the control logic usually incorporates other controlling actions in addition to those indicated. For example, many systems, particularly if multiplexing a number of separate inputs for A/D conversion, will include sample and hold units at the input. In this way the input remains constant while the conversion is made and possibly while the digital data is transferred to a separate store. The logic may also generate a 'conversion done' pulse which is used by other parts of the

Table 5.1 Comparison of A/D methods

Conversion technique	Digitsing time T_D (τ = clock rate)	Aperture time T_A (τ = clock rate)	Comments
Analog voltage to time (ramp)	$2^N \tau$	$2^N \tau$	restricted slow sampling rate
Analog voltage to frequency	$2^N \tau$	$\left(\dfrac{2^{N-1}}{3}\right)^{1/2} \tau$	as above
Feedback digital ramp	$2^N \tau$	$2^N \tau$	as above
Reversible counter	$2^N \tau$	τ	as above
Successive approximation	$N\tau$	$(N-1)\tau$	general purpose

system. Alternatively, when A/D converters are used as part of a digital transducer or voltmeter a continuous display of the output may be provided, or the output may be continually updated at a predetermined rate.

A comparison of the methods discussed above in respect of digitising and aperture times is given in Table 5.1. Some A/D units have clocks that operate up to 500 MHz and provide very high conversion rates even if high resolution is required. It should be remembered, however, that if a sample and hold circuit is incorporated in the system, or if filters are used to minimise series mode NIC at mains frequency, the overall performance is generally limited by the characteristics of these devices.

In nearly all these devices the digital outputs can, of course, be obtained in straight binary BCD or any other appropriate code.

Most complex systems will involve an amplifier, multiplexer and an A/D converter. The selection of the various operating characteristics for a combination of units does require careful consideration. For example consider a low-level signal, as from a thermocouple, which requires amplification of 1000 to bring the level to a value suitable for a ±10V A/D converter. Unless special care is taken, the noise developed by the amplifier and multiplexer could represent more than half the total output.

Consider next the multiplexer sampling at 200 kHz and an A/D converter with a 12 bit resolution, the amplifier placed after the multiplexer. This requires that the amplifier output must settle to within 1 part in about 4000 (12 bits) in less than 5 μs which is the maximum sampling period in order that a conversion can be made within an accuracy of 1 bit. At this stage it is necessary to select an amplifier having the necessary slew rate that would probably exacerbate the noise problem referred to above. It would seem that the full resolution of the A/D converter is unlikely to be of value unless special care is taken in the selection of the amplifier and low-noise multiplexer.

These are typical of the problems associated with digital systems involving A/D conversion systems.

5.5 References

DUKENFIELD, M.J. (1965): 'An introduction to statistical methods used in control system analysis', *Process Control and Automation*
PENNINGTON-RIDGE, M.D. (1967): 'A new transducer scanner for process control', *Automation,* (12), 13-16

SONNENFELDT, R.W. (1972): 'The processing of analog signals and the control of noise in digital systems'. Fifth National Power Instrumentation Symposium, Instrument Society of America, Texas

STRASSBERG, D.D. (1972): 'Multiplexing and grounding in analog-digital data acquisition systems', *Trans. Inst. Soc. Am.* **11**, 259-273

Chapter 6

Synchro/resolver conversion

6.1 Synchro systems

Before the introduction of digital systems, synchros were the most accurate and reliable shaft position analog transducers. Their widespread use in military systems led to extensive development, and synchro systems are now available that satisfy very high standards and can operate accurately over a wide range of environmental conditions, i.e. temperature, humidity, vibration, and shock.

It is therefore obvious that, with the advent of digital systems, attempts would be made to retain the basic synchro units and develop conversion techniques to provide digital outputs. This development has in fact been remarkably successful and systems are now available that match the digital encoders in resolution, absolute accuracy and dynamic response and in some ways provide a range of desirable characteristics that cannot be met by any other form of angular transducers. Many special and sophisticated techniques have been developed in control and indicating systems using synchros and many of these can be used to advantage in systems employing synchros and D/A converters. Synchros require excitation by an alternating supply and the shaft position information is usually generated as modulated a.c. signals. In most installations these signals do not need conditioning and are less vulnerable to interference from noise or ground loops than normal analog signals. The installation problems are therefore generally less severe than many other systems, including some shaft encoder installations.

Synchro pair

The simplest form of synchro system has two units, a transmitter and a receiver or transformer (Fig. 6.1). These units are essentially similar, each consisting of a two pole rotor and a stator with three windings distributed 120° apart. The rotor of the transmitter is fed with an alternating current (at 50, 60 or 400 Hz) through slip rings. The rotor winding on the transformer provides the output of the system.

The a.c. supply on the control transmitter rotor, known as the reference frequency or carrier frequency, induces signals at the same frequency across the stator leads S_1, S_2 and S_3. The line-to-line voltages V_{1-2}, V_{2-3}, and V_{3-1} have an explicit relationship to the angular position of the stator and hence the position of the input shaft with respect to the stator.

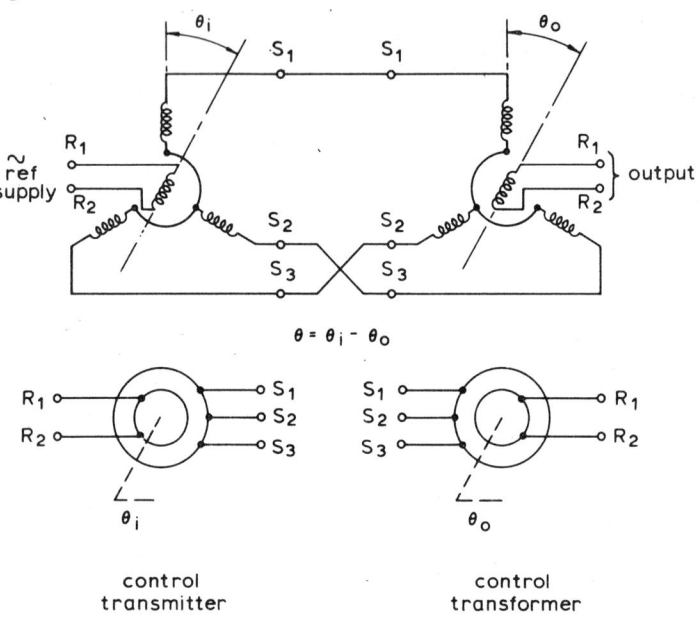

Fig. 6.1 Synchro pair

The control transformer has its stator connected directly to the transmitter as shown, and a magnetic field is set up in the transformer which is similar to that in the transmitter. The coil of the rotor in the transmitter has a voltage induced whose phase and magnitude depend on its angular position relative to the magnetic field developed by the stator coils. This voltage is at the carrier frequency and the design of the units is such that the magnitude of the signal is proportional to the sine of the angular difference between the two rotor shafts, i.e. $|V_0| \propto \sin \theta$, where $\theta = \theta_i - \theta_0$.

The waveforms in Fig. 6.2 indicate the control transformer output in relation to the reference supply. For values of θ from 0° to 90°

(shaft difference) the output is in phase with the supply and the magnitude is proportional to sin θ. Between 90° and 180°, the signal is also in phase and also proportional to the magnitude of sin θ. Thus

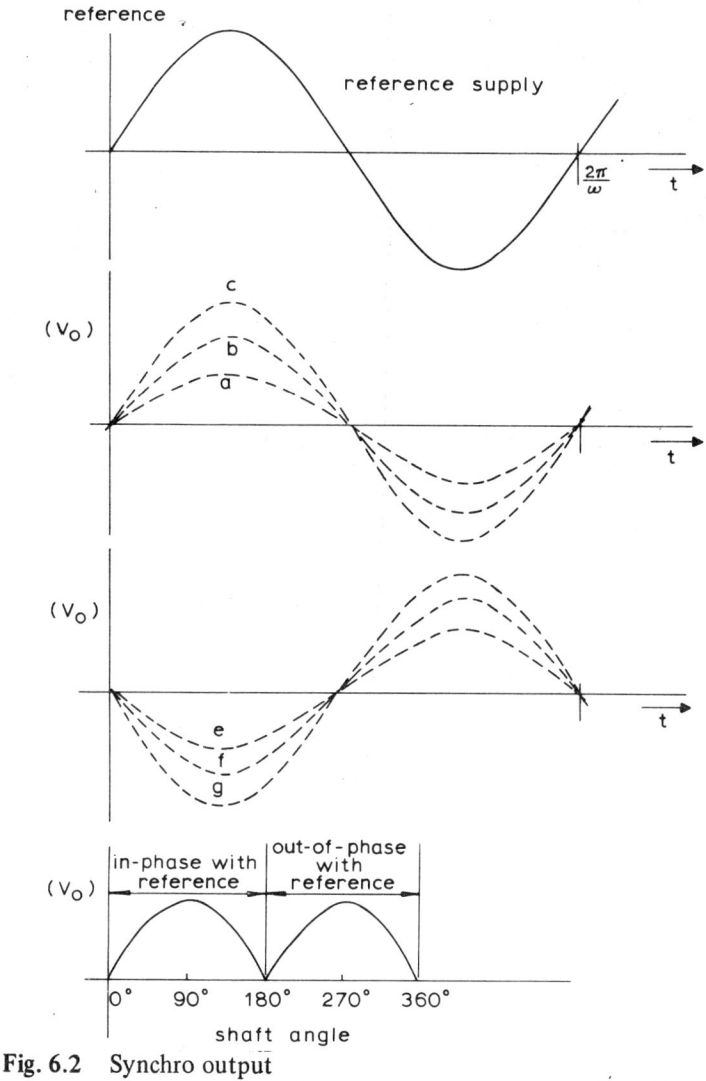

Fig. 6.2 Synchro output

Output θ : 90°–180°
$|v_a| \propto \sin\theta$
$\theta = \theta_i - \theta_0$
(a) $\theta = 30°$ and $150°$
(b) $\theta = 60°$ and $120°$
(c) $\theta = 90°$

Output θ : 180°–360°
$|v_0| \propto \sin\theta$
$\theta = \theta_i - \theta_0$
(e) $\theta = 210°$ and $330°$
(f) $\theta = 240°$ and $300°$
(g) $\theta = 270°$

there are two values of θ that give rise to a given magnitude of signal which is in phase with the reference supply. Similarly, for values of θ between 180° and 360° the output is always out-of-phase with the reference supply and its magnitude proportional to the magnitude of $\sin\theta$, with two angles for each magnitude of signal.

These units are normally used in this combination either as transducers or as input/output elements in servomechanisms, i.e. feedback control systems. The output voltage represents the error between the input and output shafts and can be used to drive a servomotor through a suitable amplifier. The connections are such that an in phase error signal would drive the output shaft towards $\theta = 0°$, i.e. the datum or null and an out-of-phase error signal towards $\theta = 360°$, that is, the datum or null.

A false zero can occur at 180° as the magnitude of the error is also zero. This is known as an unstable error since the slightest perturbation will generate an error signal to drive the output away from the 180° position and towards the zero or null position.

When used as a position transducer, the transformer shaft must be fixed, although it can be used null to set the zero datum. The output signal would normally be fed to a phase-sensitive detector together with the reference supply. The output will then be a d.c. signal, the voltage being proportional to $\sin\theta$ and the sign positive, or negative depending on whether the transformer output signal is in phase or out of phase with the reference supply. However, the value of θ must not be allowed to exceed $\pm 90°$ of the datum if ambiguity is to be avoided, since the magnitude of the transmitter output is maximum at $\pm 90°$ and then decreases again to zero at 180°.

Control differential

This unit has a three coil stator and a three coil rotor. Its function is to change line-to-line voltages so that the output voltages from the

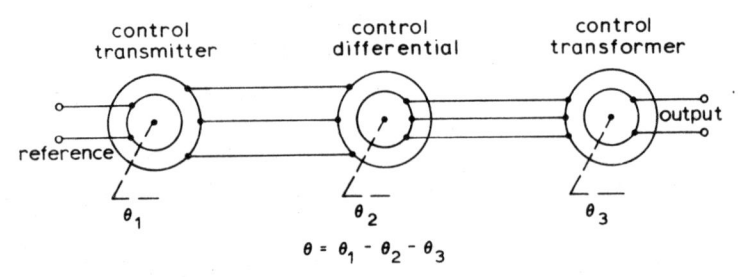

Fig. 6.3 Control differential

rotor represent the difference between the angle represented by the stator line-to-line voltage and the angular position of the control-differential rotor. In a typical installation the control differential is interposed between a control transmitter and a control transformer, as shown in Fig. 6.3.

Resolver

This unit has a rotor similar to a control transmitter and is fed by the reference supply. The stator has two windings only, and they are positioned so that one produces a voltage, at reference frequency, whose magnitude is proportional to the sine of the shaft angle, and, the second, a voltage whose magnitude is proportional to the cosine of the shaft angle.

Torque units

All the above units are control units and are designed to produce output voltages, at the reference frequency, whose magnitude is proportional to a shaft position or the difference between two shaft positions. Alternatively, the receivers can be designed to rotate the shaft to the same position as the transmitter. To generate sufficient torque, the rotor windings of both the transmitter and the receiver unit are supplied with the reference supply.

Synchro components are sometimes identified by code letters, as follows:

Unit function	Control unit	Torque unit
Transmitters	CX	TX
Differential transmitters	CDX	TDX
Control transformers	CT	-
Torque receivers	-	TR
Resolvers	RS	-

A fuller code is used for synchro units manufactured to military specifications. Additional letters represent the reference frequency, reference voltage (26V, 110V etc.) and also the physical size, which ranges from a minimum of about 12·5 mm diameter.

Brushless synchros

A further development has led to brushless synchros in which connections to the rotor are not made through brushes. This avoids the difficulties arising with brushes in any mechanism, namely effects of the environment, particularly vibration, atmospheric corrosion etc. The energy and information is transferred between a second rotor, on the

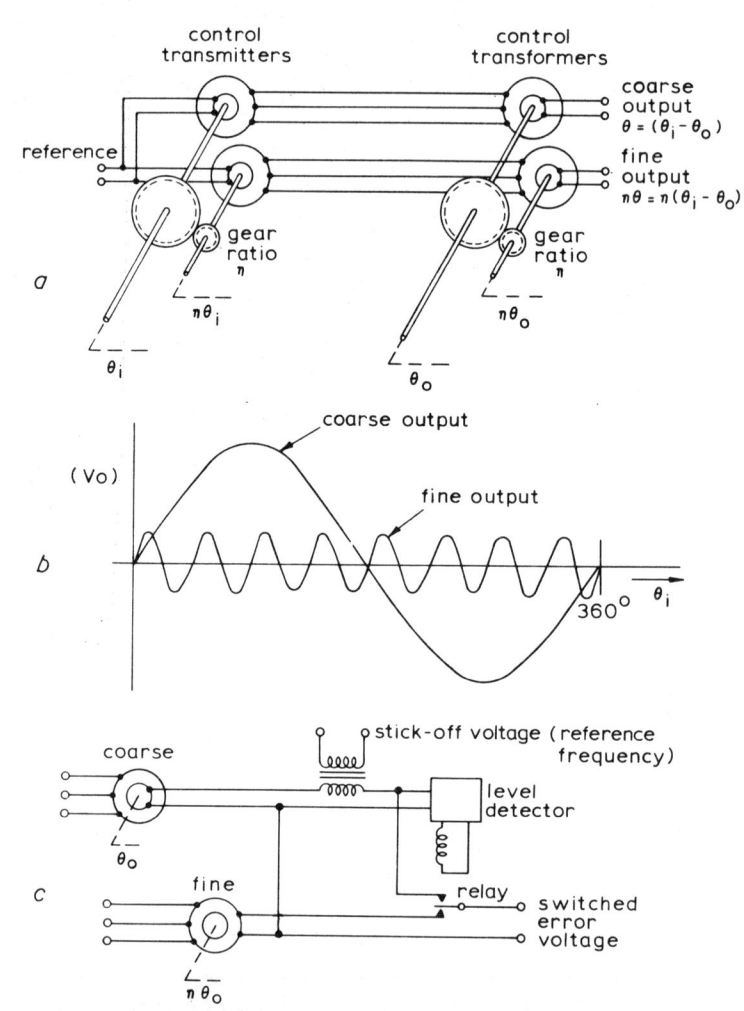

Fig. 6.4 Two-speed synchro system
    a Physical arrangement
    b Output voltages
    c Switching circuit

rotating shaft, and a second stator by a transformer action with a fixed ratio (1:1) at all shaft positions.

Two-speed synchro system

Multi-speed synchro systems consist of two (and possibly more) synchros geared together with the gear ratio n usually in the range 8 to 36. A typical system employing two control transmitters and control transformers is shown in Fig. 6.4a. The input shaft drives one transmitter rotor directly and the second transmitter rotor through a step-up gear drive so that the second rotor rotates n revolutions for one revolution of the directly driven rotor. Similarly, the output shaft Θ_0 drives one transformer rotor directly and the second through gearing having the ratio $1 : n$, as shown.

Assuming that, initially, the input and output shafts are in the same positions, one rotation of the input shaft θ_i (with the output shaft fixed) will give one complete cycle of output voltage on the coarse system and n cycles of output on the fine system. These outputs are shown in Fig. 6.4b. Remember that each output is in fact a modulated signal at the carrier frequency, which is not shown.

The overall advantage is that the coarse output gives position information over the whole 360° of rotation but provides a resolution limited to that given by the synchro. This limit can be improved by switching to the fine synchro output which, at this displacement, gives an output greater than the coarse unit since it has rotated n times the coarse unit. For example, if the smallest measurable error is 0·05° on the coarse unit, this represents n x 0·05° on the fine unit, and the measurable error is theoretically improved by a factor of n, i.e. to 0·05/n degrees of arc.

The change-over from the coarse to the fine transformer outputs can be accomplished by a level detector connected to the coarse output. This output is used as the error signal until it falls to a level that represents an angular displacement of the fine control transformer of less than 90° from zero (i.e. 90/n° of the coarse transformer) in order to prevent false nulls on the fine output.

It is easy to verify that if n is odd, unstable nulls arise on both the coarse and fine outputs at 180°. However, if the gear ratio n is even, the fine output will have a stable null at 180° which could give rise to a null at 180° position error. This can be avoided by introducing an additional 'stick off' voltage to the coarse/fine switching network, as shown in Fig. 6.4c.

6.2. Tracking converters

The simplest conversion systems use a synchro pair as a transducer and a phase sensitive detector to give a d.c. output signal. This is then converted to a digital signal by one of the techniques described in Chapter 5. The d.c. signal follows the synchro signal without any delay. In fact the only delays in the system are those normally associated with A/D converters. Owing to the ambiguity of large angle differences the error should not exceed $90°$.

However, all the information concerning the angular position of the rotor of a synchro control transmitter, relative to the stator coils, is contained in the three stator leads. This must be so as the data can be recovered in a control receiver. Alternatively, a resolver output also contains all the necessary data.

It is a logical development, therefore, to use techniques which convert the transmitter or resolver outputs directly to a digital signal representing the angular position of the shaft. The remaining parts of this chapter describe some of the techniques which are available commercially, or have been proposed.

Resolver format

Many of the methods use the resolver format as the input information. The normal 3-phase stator output of a transmitter or a differential unit can be converted to resolver format by using the Scott 'T' transformer, shown in Fig. 6.5. Remember that the primary is in fact a 3-phase signal at the reference frequency, and the outputs will similarly be signal at

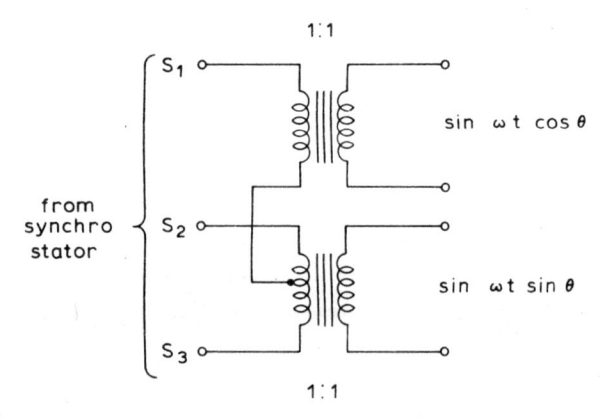

Fig. 6.5 Scott T transformer

the same frequency, and it is the magnitudes of the secondaries which are proportional to sin θ and cos θ.

For a reference or carrier voltage v_c and frequency of ω rad/s, the outputs in resolver format will be:

$$v_1 = K_1 v_c \sin(\omega t + a_1)\sin\theta \qquad (6.1)$$

$$v_2 = K_2 v_c \sin(\omega t + a_2)\cos\theta \qquad (6.2)$$

Where K_1 K_2 are constants, and a_1 and a_2 are time-phase shifts in the carrier signal caused by imperfections in the synchro components and/or the Scott transformer. Remember also that other errors can arise, derived from variations in the amplitude of the reference signal, and, in some instances, variations in frequency. The harmonic and quadrature components arising in these electromatic systems are also sources of error that give inaccurate values of v_1 and v_2 and affect accurate conversion to digital outputs.

The constants K_1 and K_2 are the effective transformer ratios between the rotor and the stator, and ideally they are equal. Also, for an ideal system, the time-phase shifts a_1 and a_2 are zero and can be ignored in most systems. Many of the other sources of error can also be minimised by special consideration of converter circuit design.

In the ideal case, the resolver outputs (or the outputs from a Scott transformer) can therefore be written as:

$$v_1 = \sin\omega t \sin\theta \qquad (6.3)$$

$$v_2 = \sin\omega t \cos\theta \qquad (6.4)$$

Phase-shift converters

This type of converter provides a relatively cheap and self-contained unit that has many applications. It has limited accuracy and suffers from certain limited operating conditions. However, because it does not necessarily need an external reference supply, it is of particular value where only a few transducers are used in a system.

The method requires the generation of a signal, at the reference frequency, which is phase shifted with respect to the reference by an angle equal to the rotor shaft position. This is achieved by the circuit shown in Fig. 6.6. If the values of R and C are chosen so that $\omega RC = 1$, the voltage v_a becomes:

$$v_a = \sin \omega t \cos \theta + \sin (\omega t + \pi/2) \sin \theta \qquad (6.5)$$

$$v_a = \sin \omega t \cos \theta + \sin \omega t \cos \pi/2 \sin \theta + \cos \omega t \sin \pi/2 \sin \theta$$

$$v_a = \sin \omega t \cos \theta + \cos \omega t \sin \theta$$

$$v_a = \sin (\omega t + \theta) \qquad (6.6)$$

By using Schmitt trigger circuits, the reference sine wave and v_a can be squared and used to produce a start-stop pulse for a counter, as shown

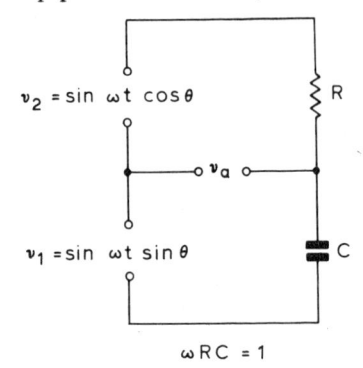

Fig. 6.6 Phase-shift converter

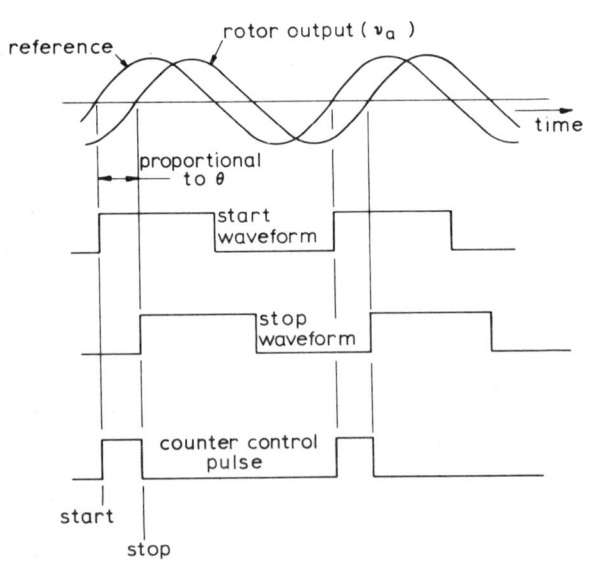

Fig. 6.7 Phase-shift encoder waveforms

in Fig. 6.7. As the duration of this pulse is proportional to θ, the counter total represents the angle θ. The counter can be up-dated at the reference frequency as the count takes less time than time for one cycle.

A similar system* is shown in Fig. 6.8. A crystal oscillator operating at 360 kHz is divided by 360 and squared to produce a 1 Hz square wave. This is used to generate the start of a counter control pulse. It is also shaped to produce a sine wave, harmonic free, which is fed to the stator windings of the resolver. This generates a 1 kHz circular field in the resolver.

The resolver output is the rotor winding which develops a constant amplitude waveform that is phase shifted relative to the 1 kHz stator

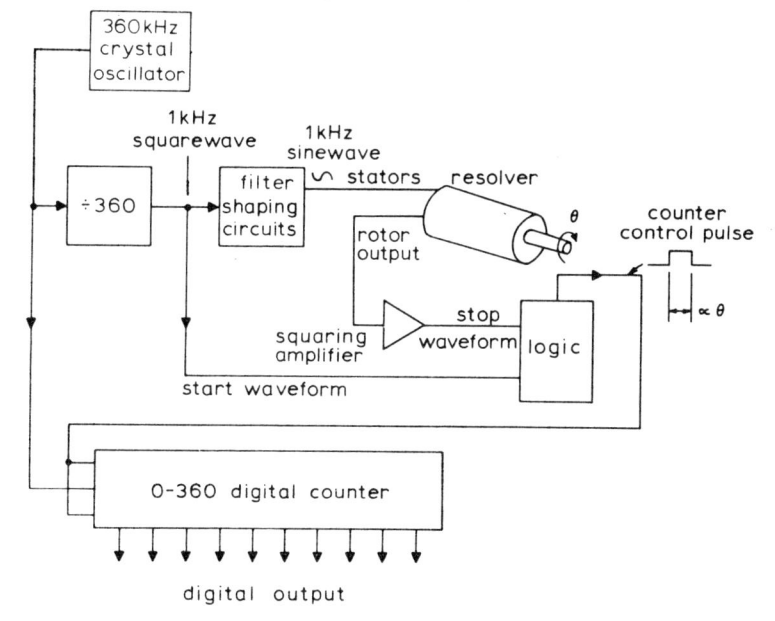

Fig. 6.8 Resolver phase-shift encoder

supply. This output is squared in a Schmitt trigger squaring circuit to give a stop signal to the counter control pulse.

This pulse has a maximum length of 1 ms which represents an angular displacement of the rotor shaft of 360°, i.e. one revolution. For this time 360 pulses are counted into the counter. Thus, the digital output has a total of not less than 9 bits and the overall resolution is only 1 in 360, i.e. 1°, which is often sufficient. This can, of course, be increased by using two resolvers in a two-speed synchro

*manufactured by Muirhead Ltd.

system. Increasing the resolution by increasing the clock frequency is not usually economic, largely due to other sources of error. There are the difficulties of maintaining $\omega RC = 1$, due to variations caused by temperature changes, and also phase errors in the resolver arising primarily from temperature changes.

In the system described above, the maximum speed of operation is limited to about 15 rev/min, if the resolution of 1 in 360 is to be maintained. This arises from the slow updating rate, i.e. the counter is updated at the reference supply frequency 1 kHz. Hence, for an accurate output, a change of $1°$ must not occur within 1 ms, i.e. the maximum change is $1000°/s$ or $(1000 \times 60)/360 = 15$ rev/min.

Other units working on this principle are available with greater resolution and are not so sensitive to temperature changes, although a limitation on the maximum speed of operation still applies.

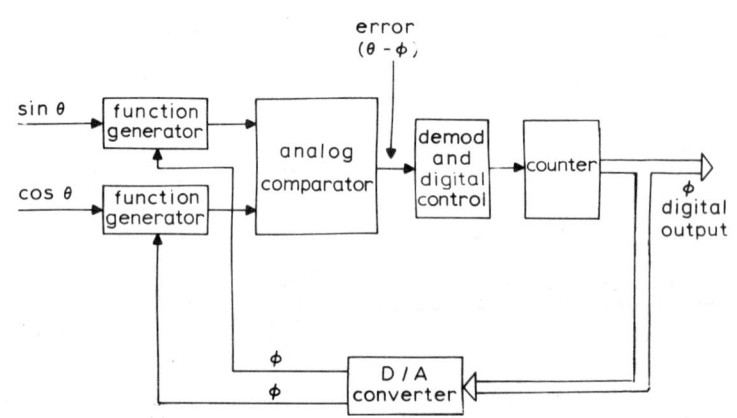

Function-generator converters

These systems use a feedback technique in which the digital output is fedback to function generators which develop signals that are used to drive the digital output towards a value equal to the shaft position of the resolver. An equilibrium position is reached whenever the output is equal to the shaft position.

The principle is shown in Fig. 6.9. The input signals in resolver format, which represent the rotor shaft position θ, are fed into function generator circuits to which another signal is applied representing the angle ϕ, defined by the digital output. The function generator outputs are fed into a comparing circuit which develops a signal that is

Fig. 6.9 Function generator converter

a function of $(\theta-\phi)$, i.e. the error between the shaft position and the angular equivalent of the digital output.

This error signal is used to drive a gate and/or counting circuit which changes the digital output, varying the value of ϕ so as to null or zero the value of $(\theta-\phi)$. The digital value of the output ϕ will then be equal to the shaft position.

This scheme is more expensive than the phase-shift converter described earlier, but the use of integrated circuits has reduced the cost difference. Also, the overall accuracies are better, and performances of ±2 seconds of arc are attainable.

Other advantages are:

(i) The system is inherently a ratio technique as it depends on the ratio of the two resolver signals. From eqns 6.3 and 6.4,

$$\frac{v_1}{v_2} = \frac{\sin \omega t \sin \theta}{\sin \omega t \cos \theta} = \tan \phi \qquad (6.7)$$

This makes it easier to design the system to be less sensitive to harmonics in v_1 and v_2 and also unwanted quadrature components.

(i) The system gives a continuous real-time measurement of ϕ limited only by the speed of generation of the digital output.

(iii) It is almost independent of the reference supply frequency and will tolerate considerable variations.

Actual systems working on this principle take a variety of forms depending on the type of function-generator used, the manner in which the error signal $(\theta-\phi)$ is developed and, finally, how this signal is used to generate the digital output signal. Some of the systems are, in effect, 'tracking converters' which are described in more detail in the next section. The basic difference between a straight function-generator converter and a true tracking converter is the technique used to develop the digital signal from the function-generator outputs.

The function-generators used in commercial systems are effectively hybrid multipliers which generate an analog output signal which is the product of the analog input and a function of a digital input.

The digital input represents the output angle ϕ, and the functions of this angle used in the multiplier are $\sin \phi$ and $\cos \phi$. The individual function generator outputs are as follows:

$$v_a = v_1 \cos \phi \qquad (6.8)$$

and
$$v_b = v_2 \sin \phi \tag{6.9}$$

From eqns. 6.3 and 6.4

$$v_a = \sin \omega t \sin \theta \cos \phi \tag{6.10}$$

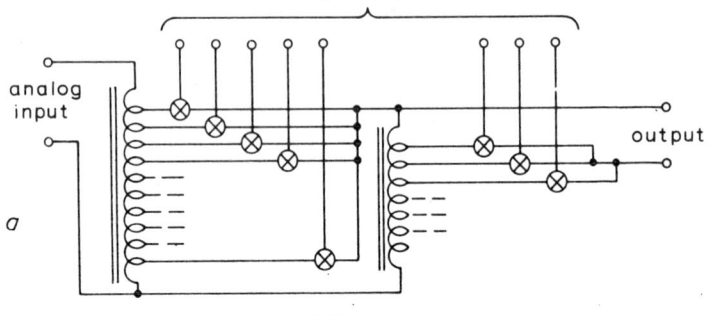

$$v_b = \sin \omega t \cos \theta \sin \phi \tag{6.11}$$

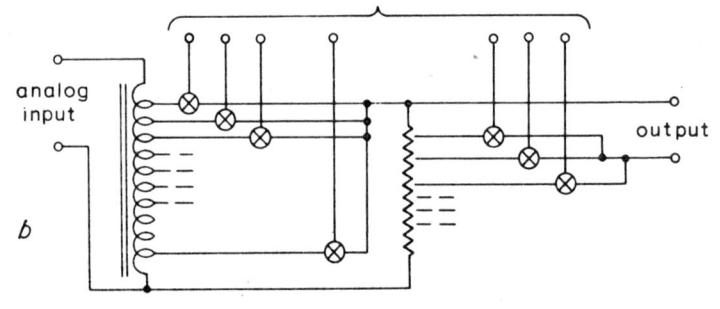

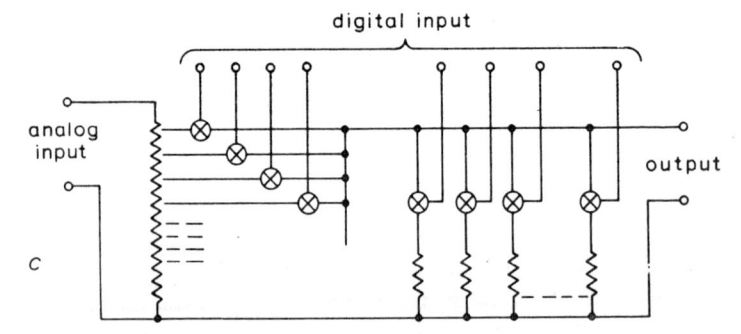

Fig. 6.10 Function generator principles
    a Tapped-ratio transformer
    b Transformer resistor network
    c Linear resistor network

These two outputs are then fed to a differential amplifier to find the difference or error signal v_e:

$$v_e = v_a - v_b = \sin \omega t \, (\sin \theta \cos \phi - \cos \theta \sin \phi) \qquad (6.12)$$

$$v_e = \sin \omega t \sin (\theta - \phi) \qquad (6.13)$$

This is an alternating voltage at the reference frequency whose amplitude at any instance represents the sine of the angle difference between the shaft position θ and the angular equivalent of the digital output.

Several different types of function-generators are in use and some of these are shown in Fig. 6.10. The multitapped auto-transformer shown in Fig. 6.10a is used by some manufacturers and represents a stable and precise way of obtaining the required sine or cosine function of the digital signal representing the angle ϕ.

The ratios of the various taps are arranged to give the required function and these are switched into the output circuit by solid-state switches, similar to the switches used in multiplexers. As the transformer source is one of the resolver outputs (i.e. either eqn. 6.3 or 6.4), the output of the function generator will be as given by eqns. 6.10 or 6.11.

One of the serious drawbacks of this unit is the switching transients that can occur with the inductive circuits. The arrangement shown in Fig. 6.10b overcomes these effects because the auto-transformer has been replaced by a resistor network. The resistor values are chosen to give the same overall function as the transformer system.

The resistors must be high stability and high precision if the same degree of accuracy and stability is to be achieved as given by the transformer units. They are, however, much cheaper and therefore many commercial units use resistor networks.

The arrangement shown in Fig. 6.10c uses a linear resistor network and achieves its sine or cosine function by selective loading of the ratio divider. This sytem provides the most economical units, being cheaper and lighter than the other systems.

The switching arrangements necessary for all three of these techniques are more complicated than the illustrations suggest.

As explained earlier, the outputs of the two function generators can be fed to a differential amplifier to generate the waveform described by eqn. 6.13. Demodulation gives the output $\sin (\theta - \phi)$ which can be converted to a digital signal by conventional techniques. The digital format may be in any desired code, but it should be noted that the function generators must be capable of interpreting this digital information to the sine/cosine functions required.

The method usually used by manufacturers is to use the demodulated error signal to drive an oscillator system to develop pulses that can be counted into the output counter until the value of ϕ, the output angle, equals the resolver shaft position θ. This type of converter is often described as a 'tracking converter'.

Tracking converters

These converters use a more sophisticated error processor and digital generator than the function-generator systems described above, although the determination of a demodulated and smoothed error signal proportional to sine $(\theta - \phi)$ is still similar. As shown in Fig. 6.11, this

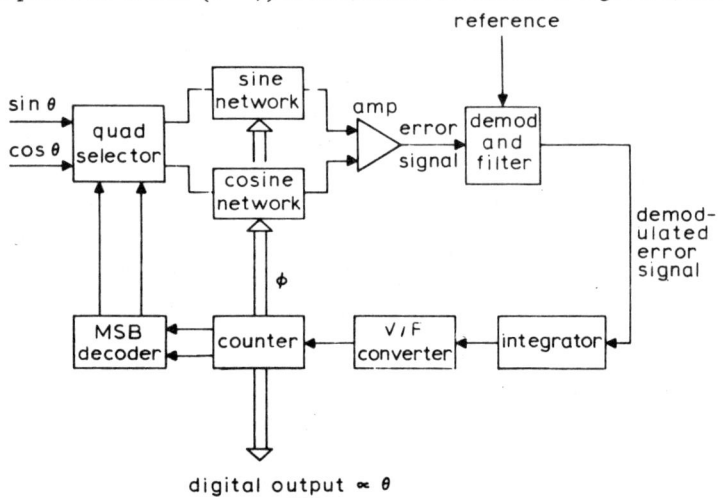

Fig. 6.11 Tracking converter

error signal is passed to an analog integrator so that its output is the time integral of the error signal. This output then drives a voltage/frequency (V/F) converter, similar to those described in Chapter 3, which gives rise to a train of pulses whose frequency is proportional to the integrator output.

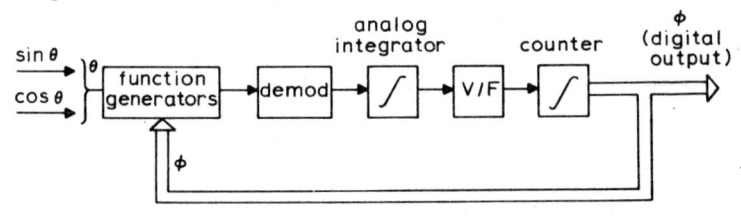

Fig. 6.12 Tracking converter block diagram

At the same time, the sense of the error [e.g. ($\theta - \phi$) negative, or ($\theta - \phi$) positive] is used to generate a count-up or count-down control signal to the counter which accumulates the V/F converter pulses. The digital number in the counter is, functionally, the time integral of the V/F converter output, and is an incremental integrator.

Thus, the overall system can be represented by a block diagram corresponding to a feedback control system with two integrators in the forward path, and in feedback terminology this would be known as a type 2 servomechanism (Fig. 6.12). The significance of this is found in the inherent performance characteristics which a type 2 servo has. Briefly, the important characteristics can be summarised by its response to different types of inputs:

(a) if the input is stationary then the output Φ will be identically equal to θ

(b) if the input is moving with steady angular velocity (i.e. the rotor shaft is rotating) then, again, the output changes and thus represents the same angular velocity. At any instant the output will be the same as the input (i.e. ϕ will be identical to θ)

(c) if the input is accelerating (i.e. the velocity of θ is changing) then the output will have the same acceleration, but at any instant the angle ϕ will lag behind the angle θ.

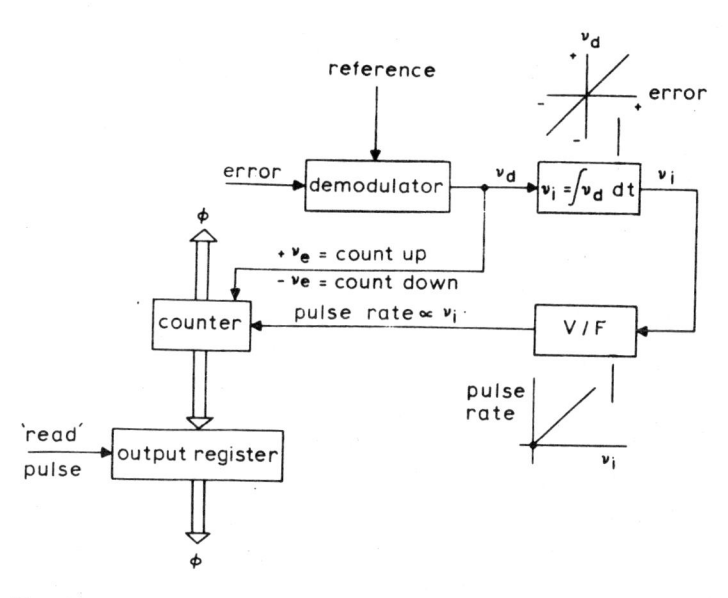

Fig. 6.13 Tracking converter control

It is the particular characteristics of zero velocity error (*b* above), inherent in this type of tracking converter, that makes them the most popular of all the types of tracking converter. Many other systems do not have this zero velocity error.

Fig. 6.13 shows further details of a system of this type. The error signal developed from the function generators is used to update the counter. After demodulation and smoothing, a d.c. signal is developed, which is directly proportional to the error, having a positive sign if θ is greater than ϕ and negative sign if θ is less than ϕ. An output from the demodulator is fed to the counter for a count-up ($\theta > \phi$) or count-down (for $\theta < \phi$).

The d.c. error signal v_d is also passed to a simple integrator to provide the time integral v_i that is used by the voltage/frequency converter and pulse shaper to give an output of pulses whose pulse rate (or frequency) is directly proportional to v_i. These pulses are added to or subtracted from the counter until the error signal is reduced to zero, i.e. when $\theta = \phi$. Note that, for a constant error, the pulse rate is rising as the time integral v_i of a constant error v_d is increasing.

Consider the case in which the resolver rotor is rotating at constant velocity. Any error ($\theta - \phi$) in the system gives an increasing value of v_i. This rises until the pulse rate is sufficient to maintain a change in ϕ which just equals the change in θ due to its constant velocity. Thus the digital output ϕ will follow the input θ, providing the velocity of θ is not larger than the equivalent maximum pulse rate that can be developed by the V/F converter.

During periods of acceleration and deceleration of θ, as when starting, stopping or changing speed, small errors arise. The maximum pulse rate from the V/F converter also decides the minimum time in which the system can give a correct output from the maximum error position, i.e. $\pm 180°$.

The introduction of the phase-sensitive demodulator has the advantage that it reduces the effects of quadrature components on the resolver or synchro outputs. These unwanted components arise particularly when the rotor is rotating as it tends to act as a generator and the quadrature components can be alarmingly high relative to the true signal voltages at high rotational speeds. Hence, systems that are required to operate and provide accurate data at high angular velocities should minimise the quadrature effects. This can be achieved with careful demodulator design.

In the system shown in Fig. 6.11, the two most significant bits of the output are used in a quadrant selection network to provide sin θ and cos θ signals to the function generators so that ($\theta - \phi$) is always less than 90°. Fig. 6.14a shows that for angles less than 90°, i.e. the first quadrant, both sine and cosine are positive. In other quadrants, the signs change as shown. The quadrant selector network accepts the two most significant bits from the counter and switches the resolver inputs, unchanged or their inverse, to the function generators as sine and cosine functions of the first quadrant only.

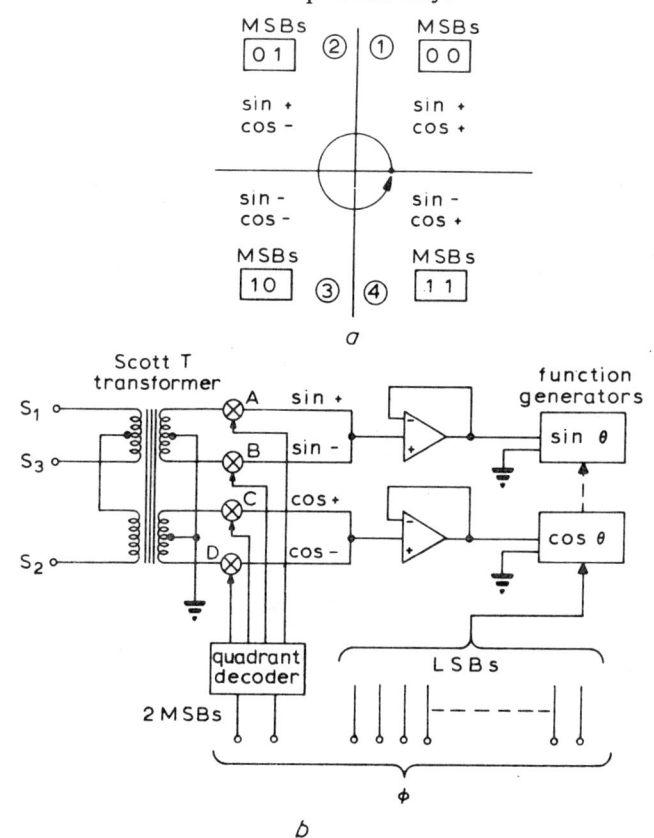

Fig. 6.14 Quadrant switching

    a   Quadrant coding   *b*   Switching circuits

		Switches				Signals to generators	
MSBs	Quadrant	A	B	C	D	sin	cos
00	1	on	off	on	off	normal	inverted
01	2	on	off	off	on	normal	inverted
10	3	off	on	off	on	inverted	inverted
11	4	off	on	on	off	inverted	normal

The switching network is shown in Fig. 6.14b. A digital switching system is driven by the two MSBs, representing the angle ϕ and switches the outputs of the transformers into the function generators according to the requirements of the table shown. The transformer primaries may, of course, be connected as a Scott-T transformer to accept the three lines from a synchro transmitter stator.

The type of tracking encoder described above is available commercially to provide a resolution up to 14 bits, representing 1·3 minutes of arc, with an overall accuracy of about ±4 minutes of arc. The major contribution (±3 minutes) to the error generally arises in the function generators. Other errors, arise from the Scott-T transformer (±1 minute) and drift due to temperature variations.

High tracking speeds, i.e. the maximum speed at which the above accuracy is maintained, can be up to about 1500° per second (240 rev/min) for a 14 bit resolution. Generally, faster tracking speeds are possible with converters having low resolution, and lower maximum speeds are achieved with higher resolution units. All these are considerably faster than phase-shift encoders although the response to a step input is slower. A typical tracking encoder could take up to 0·5 s to accurately measure a step change of 180° due to the maximum pulse rate obtainable from the V/F converter. The phase-shift encoder described earlier, would take only 0·5 ms to accurately measure a step change of 180°.

High-resolution tracking converters

The resolution of a shaft position system using tracking converters can be increased by the use of a two-speed system. While two separate synchro digital converters could be used, the usual arrangement is to economise by using a common counter and error processor.

A typical arrangement is shown in Fig. 6.15. Both the course and fine synchros have their own quadrant selectors and function generators etc. The analog error of each channel is connected to a crossover detector. For large errors the crossover detector allows the error signal from the coarse system to be fed to the error processor, i.e. the integrator and voltage/frequency converter. If the coarse error is less than that represented by ±90° of the fine system, the crossover detector will switch the fine synchro error to the error processor. Note that the multipliers of both circuits are presented simultaneously with the digital counter output. This ensures a smooth changeover from coarse to fine and vice versa. The crossover level depends on the coarse/fine gear ratio of the synchros.

This technique can give a high resolution without any increase in the fundamental accuracy of the converters. It does, however, require additional synchro units and a more complex converter unit.

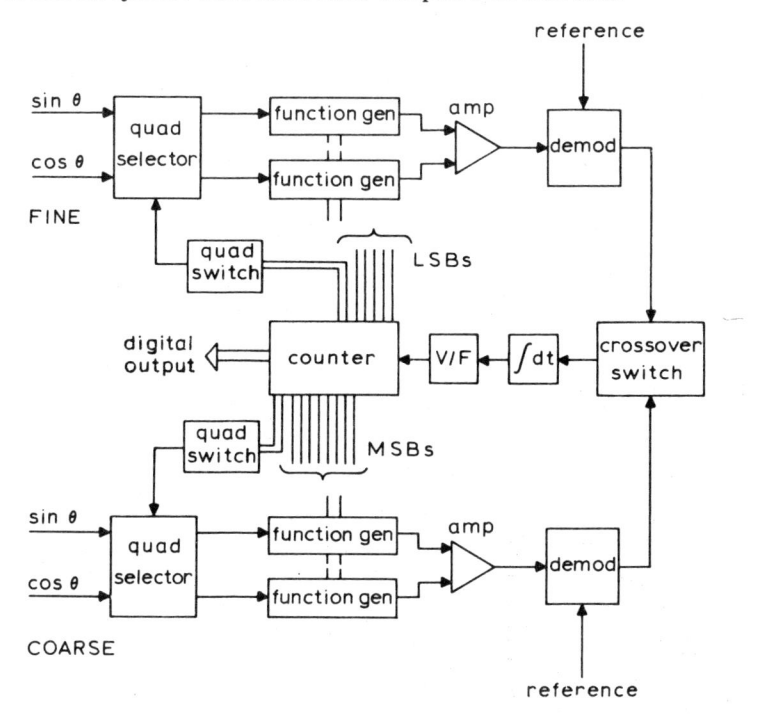

Fig. 6.15 Two-speed tracking converter

An alternative technique provides a greater resolution from a single resolver format input.* In principle, the system is similar to the normal tracking converter but has two error networks which together control the error processor. As will be seen in Fig. 6.16, the input of the system is similar to the tracking synchro already described, but additional outputs are developed by the function generators which are summed in a second summing amplifier.

The digital angle ϕ is considered to be split into two parts, ϕ_1 represented by the six most significant bits, and ϕ_2 by the least significant bits. For example, consider a converter with a resolution of 16 bits: ϕ_2 will be ten bits and will represent a maximum angle of 5° 37·5 min. That is 1/64 of 360° since the least significant bit, of the six most significant bits, represents 1/64 of a complete revolution.

*marketed by Muirhead Ltd.

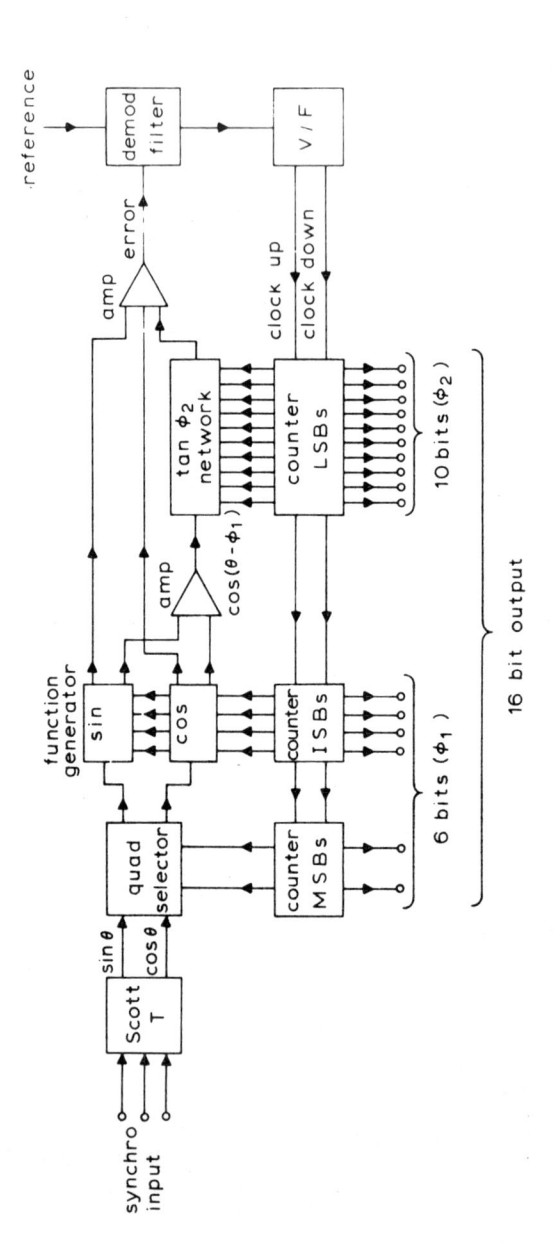

Fig. 6.16 16-bit synchro converter

Synchro/resolver conversion

The function generators produce two outputs each. Omitting the carrier signal terms ($\sin \omega t$), these are

$$\sin \theta \cos \phi_1 \qquad (6.14)$$

$$\cos \theta \sin \phi_1 \qquad (6.15)$$

as produced in the usual tracking converter, and two additional signals

$$\cos \theta \cos \phi_1 \qquad (6.16)$$

$$\sin \theta \sin \phi_1 \qquad (6.17)$$

The first two signals are fed to an amplifier to produce

$$(\sin\theta\cos\phi_1 - \cos\theta\sin\phi_1) = \sin(\theta - \phi_1) \qquad (6.18)$$

again as in the conventional tracking converter. The second two are taken to a second amplifier to produce

$$(\cos\theta\cos\phi_1 + \sin\theta\sin\phi_1) = \cos(\theta - \phi_1) \qquad (6.19)$$

The output then passes to an additional function generator which multiplies the input by $-\tan \phi_2$ giving rise to an output

$$-\cos(\theta - \phi_1) \tan \phi_2 \qquad (6.20)$$

This is added to the signal given by eqn. 6.18 by an additional input to the first amplifier. Thus, the final error signal will be given by

$$\text{error} = \sin(\theta - \phi_1) - \cos(\theta - \phi_1)\tan \phi_2 \qquad (6.21)$$

The error signal to the error processor will therefore be zero only when

$$\sin(\theta - \phi_1) = \cos(\theta - \phi_1) \tan \phi_2 \qquad (6.22)$$

When this is achieved the total digital signal will be made up of that representing ϕ_1, and that representing ϕ_2 in the above relationship.

Function generator inaccuracies limit the overall resolution, which is at present practicable with this system, to 18 bits, representing about 5 seconds of arc.

6.3 Sampling converters

As described earlier, the data needed to determine the shaft position from the output of resolver or a synchro/Scott transformer system are contained in the amplitude of sine and cosine signals. In fact, as the carriers of the sine and cosine signals are of the same phase (except for rotor-stator phase shift) it is only necessary to measure the peak amplitudes of each signal simultaneously. The resultant signals can be stored as d.c. signals and subsequently processed to develop a digital signal proportional to the actual shaft angle.

This mode of operation, i.e. a separate sample-and-hold system for each of the resolvers, enables one converter to be used for a number of different resolvers or synchros. The output of each sample and hold circuit is sampled in turn, or on command, using techniques similar to those used for multiplexers (Chapter 5). The converter output is connected to the indicator or computer as required. An alternative method is to provide each synchro or resolver with its own demodulators, and multiplex the d.c. outputs into a single converter.

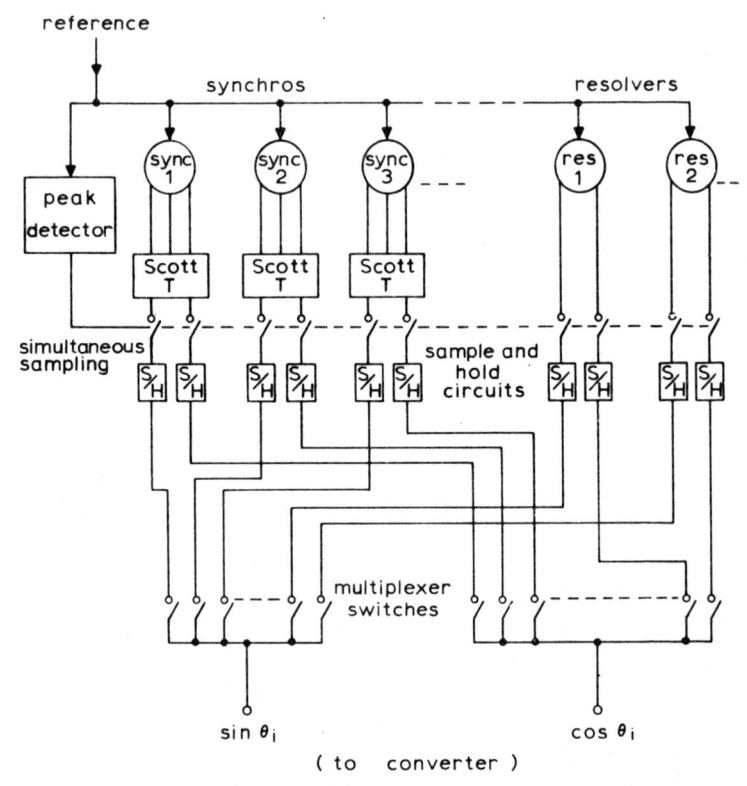

Fig. 6.17 Synchro/resolver sampling

Sampling techniques

An example of the first sampling technique is illustated in Fig. 6.17, which shows the elements required. These are a carrier peak-detector with two sample-and-hold circuits for each channel to be multiplexed. The outputs of the synchros or resolvers can then be multiplexed in sequence, as d.c. signals, to the converter. Normally, two samples are taken for each cycle of the carrier wave, one at the positive peak and another at the negative peak.

Thus, the sampled data from each channel is updated twice every cycle or once every 10 ms for a 50 Hz carrier, and once every 1·25 ms for a 400 Hz carrier.

The multiplexer is usually arranged to sample all channels during one half cycle of the carrier wave which means that the actual digital signal from a particular synchro could be delayed by up to 10 ms for 50 Hz carrier, or 1·25 ms for a 400 Hz carrier. For a shaft input revolving at 10 rev/min this delay could represent an angular error of 0·6° or 0·075°, respectively. This error becomes greater if a single pair of sample and hold circuits are used and the multiplexing is done on the resolver formatted signals. Sampling at every peak of the carrier waves would mean only two channels would be sampled in each cycle and for n channels the delay in updating the signal would be $n/2f$ seconds where f = carrier frequency in hertz. These sampling techniques are therefore restricted to slowly moving shaft positions if accuracy is to be maintained.

An alternative to the simultaneous sampling method is to provide each synchro or resolver with its own tracking converter which, as explained earlier, maintains an accurate and continuous digital output of

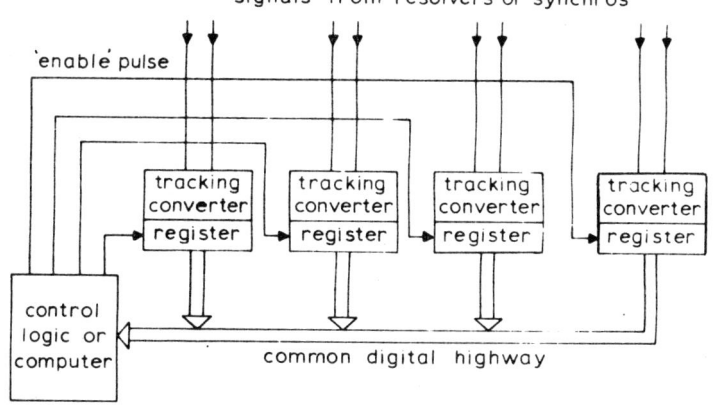

Fig. 6.18 Computer-controlled tracking converters

the input shaft position even at comparatively high rotor speeds. As shown in Fig. 16.18, the output registers of all the converters are connected to a common data highway, and, when an appropriate 'enable' command is received, the switches on the selected register close and connect the digital outputs to this highway.

This last method is the fastest and most accurate system but is generally more expensive in both individual components and installation. However, this system does minimise some of the installation problems associated with the cheaper simultaneous sampling methods when it is necessary to transmit the d.c. sampled signals over long distances to the common converter. These problems, as described in Chapter 5, are primarily caused by noise on the d.c. transmission system. There are also other problems in the simultaneous sampling method related to the carrier wave, as it is necessary for all resolvers etc. to be supplied with the same reference. Quadrature, and harmonic errors can occur with long transmission lines and give errors in the output signals.

Successive-approximation converters

The basic principles of successive-approximation converters are similar to the successive approximation A/D converter described in Chapter 5. The circuits are adapted for the particular function of converting the two d.c. signals from the sample-and-hold circuits, representing the sine and cosine of the shaft angle, into a digital signal proportional to the shaft angle.

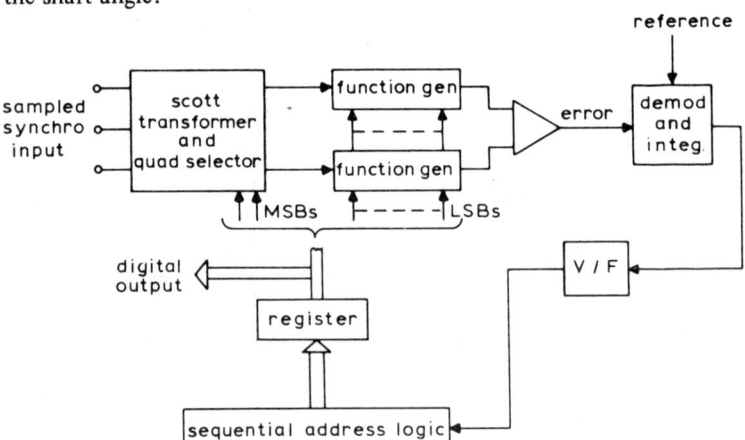

Fig. 6.19 Successive-approximation converter

The system is shown diagrammatically in Fig. 6.19, and similarity with the tracking converters described earlier can be seen. The quadrant selection and multiplier circuits are similar; the major changes are in the simplified error-processor arrangements. The error-processor is required to cause the digital output to reach the same value as the sampled input angle represented by the synchro input. The error processor sets the digital output register by the process of successive approximation of the sequential address logic. That is, selection of most significant digit first and the successive digits in order of significance, retaining those in the register if the error $\sin(\theta - \phi)$ is positive (i.e. $\theta > \phi$) and setting to logic '0' those where the $\sin(\theta - \phi)$ is negative (i.e. $\theta < \phi$).

As explained in Chapter 5, successive approximation systems can be extremely fast. Therefore, a large number of channels could be sampled and converted in each half cycle of the carrier frequency.

Harmonic-oscillator converter

The input to harmonic-oscillator converter is two d.c. signals proportional to the sine and cosine of the input angle, generated either from separate demodulators on each resolver output or outputs obtained from peak sample-and-hold circuits.

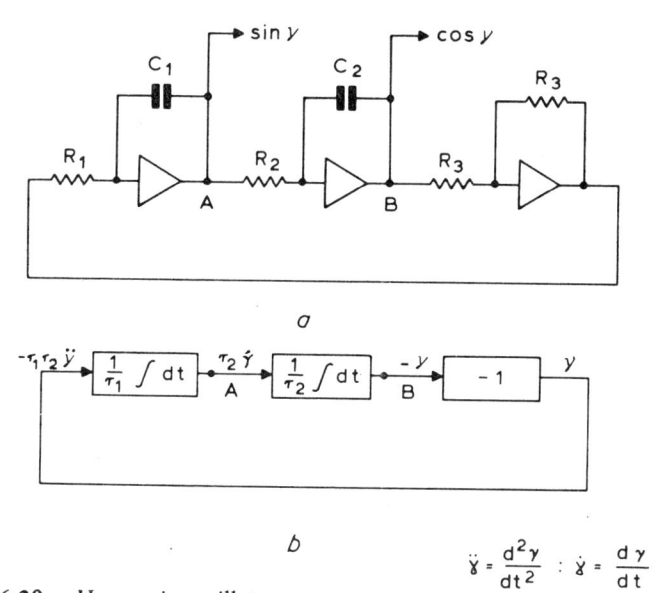

$$\ddot{y} = \frac{d^2 y}{dt^2} \quad ; \quad \dot{y} = \frac{dy}{dt}$$

Fig. 6.20 Harmonic oscillator
    a Circuit diagram   *b* Block diagram

156 Synchro/resolver conversion

The harmonic oscillator itself is basically a second-order analog circuit having two integrators and an inverter as shown in Fig. 6.20. In the conventional analysis of analog circuits the output of the inverter γ is also the input to the first integrator. Taking into consideration the negative signs resulting from the inverters and the integrators, and also the time constants τ_1 and τ_2 of the two integrators, we can equate

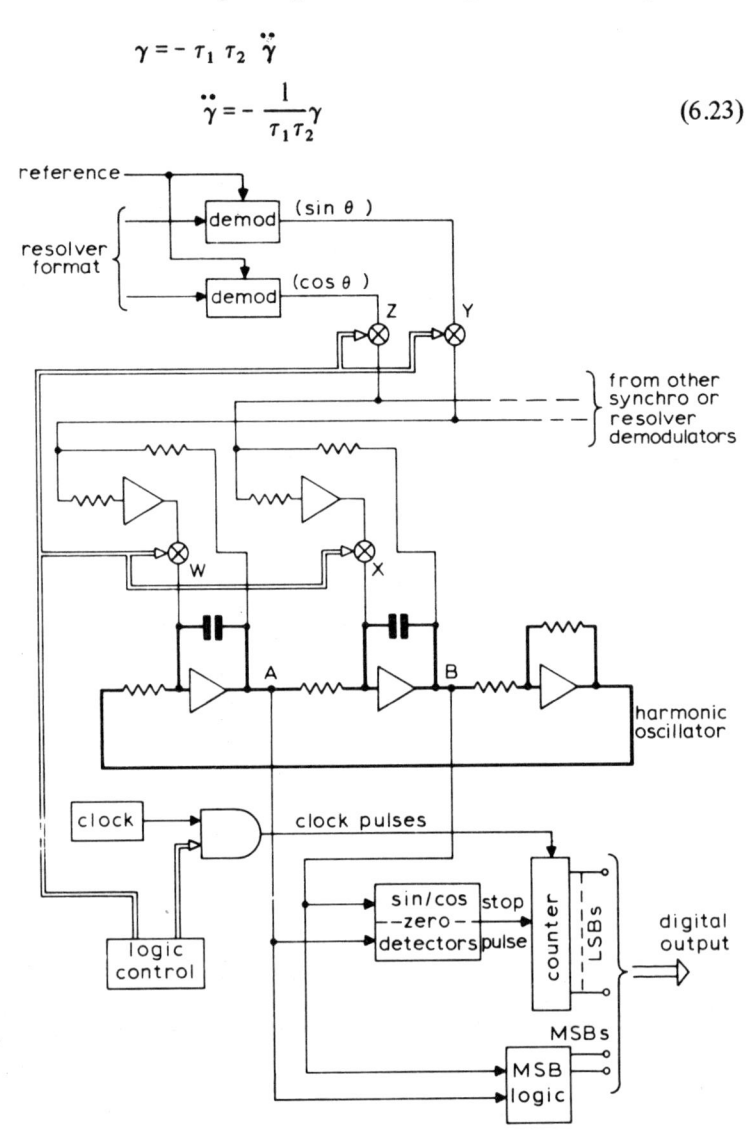

$$\gamma = -\tau_1 \tau_2 \ddot{\gamma}$$

$$\ddot{\gamma} = -\frac{1}{\tau_1 \tau_2}\gamma \tag{6.23}$$

Fig. 6.21 Harmonic oscillator converter

where $\tau_1 = R_1 C_1$ and $\tau_2 = R_2 C_2$. This is the equation of simple harmonic motion with the loop frequency ω_L given by

$$\omega_L = \frac{1}{\sqrt{\tau_1 \tau_2}}$$

and for $\tau_1 = \tau_2$

$$\omega_L = \frac{1}{\tau} \text{ rad/s} \tag{6.24}$$

or

$$F_L = \frac{\omega_L}{2\pi} = \frac{1}{2\pi\tau} \text{ Hz}$$

Relatively, the outputs at A and B will therefore be sinusoidal voltages, B being the integral of A. They can therefore be considered as the cosine and sine of the natural frequency of the oscillator. This natural frequency is usually set to about 100 or 200 Hz.

The complete converter incorporates a number of FET switches that are used to set the oscillator at predetermined values, given by the values of sine θ and cos θ and also to stop and start the harmonic oscillator. The converter includes a timing control system and a clock and also a counter to generate the digital ouput.

The complete system is shown in Fig. 6.21 and functions as follows: The oscillator is held in a non-oscillating mode by closing the FET W and X. This effectively short circuits the integrating capacitors of the two integrators. At the same time the sine and cosine values of the input angle θ are switched through closed switches Y and Z.

The oscillator is held at this state for a sufficient time to ensure that the capacitors are fully charged. This sets the initial conditions at A and B, identically equivalent to sin θ and cos θ, respectively. The programme unit then opens all four switches (W, X, Y and Z) allowing the oscillator to run at its own natural frequency ω_L, with the sine and cosine outputs starting at these initial values. The clock pulses are counted into the least significant bits of the output register, starting from the instant the oscillator commences to run until the zero crossing of either the sine or cosine outputs of the oscillator. The two most significant bits representing the quadrant are derived from the initial sine and cosine voltage polarities as explained earlier.

Typical waveforms are shown in Fig. 6.22 of the oscillator outputs and the occurrence of the start-stop signals for the clock. Shown on this diagram, but not included in the block diagram of Fig. 6.21, is a reset period which resets both integrators to zero before switching to a new set of initial conditions from the same resolver system or, through a multiplexer, from another pair of d.c. signals representing the sine and cosine of a second resolver. The reset period also allows time for the output to be strobed into other parts of the overall system and the counter to be set to zero, ready for the next count.

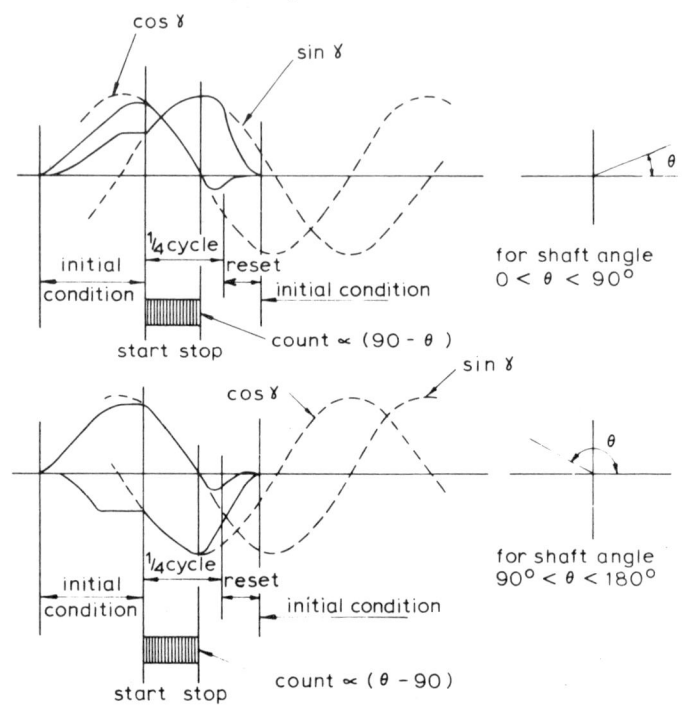

Fig. 6.22 Harmonic oscillator waveforms

For the system shown, the accumulated count is the complement of angle θ, i.e. a count representing $(90-\theta)$ will be generated in the counter. The value of θ can be determined by taking the complementtary output of the counter as the value of θ.

The count never occupies more than 1/4 of a cycle of the oscillator. Thus, allowing for reset and initial condition setting, the complete readout cycle can be achieved in less than one cycle of harmonic oscillator frequency. Thus, if the frequency of this oscillator is 200 Hz, then 200 conversions/s are possible. With a clock rate of 1 MHz, this could give outputs with a resolution of 1 in 1250 for the least significant bits only. Together with the two most significant bits derived from the sine and cosine signs in the quadrant selector, this gives a total output of 12 bits. Other combinations of clock rate and oscillator frequency will give different resolutions.

The primary accuracy of the system depends on accurate zero detectors, stable clock rate and also a stable harmonic oscillator. Oscillator stability is difficult to achieve as it depends on the RC time constant of the integrators. It is the ratio of the clock rate to the integrator frequency that usually determines the attainable accuracy and resolution. This ratio is kept constant in some systems by using a phase-locked clock generator which tracks any drift in the harmonic frequency.

Converter errors

Synchro to digital converters have the disadvantage that they are subject to the errors which invariably arise in electromagnetic transducers, particularly those associated with rotating parts, together with some of the errors that arise in conventional A/D conversion techniques.

Some of these errors have already been mentioned and discussed but are summarised below for the sake of completeness.

(a) Carrier variations
Variations in carrier frequency cause errors in phase-shift converters, particularly those with RC phase-shift circuits. Unwanted harmonics in the carrier or generated in the synchro signal can cause errors in some sampling converters since the distortion may considerably affect the peak signal value.

(b) Rotor-to-stator errors
This includes a range of possible errors due primarily to the physical character of synchros or resolvers. Rotor-to-stator time shift errors have already been discussed. Similar errors can arise from various circuit components such as transformers, quadrant selectors and asymmetric loading of synchro or resolver by the converter inputs. Quadrature voltage, as previously discussed, arises only when the rotor has an angular velocity and can be minimised only by careful selection of the converters. Errors caused by running at greater speeds than the converter and output lags arising when the shaft accelerates, can be minimised only by correct selection of the converter to suit specific requirements.

(c) Environment
Changes in performance caused by shock, vibration, ambient temperature changes, aging, humidity and atmospheric pressure etc. vary according to the design techniques employed. Naturally, those devices

(d) Signal errors

This includes all those errors which arise as a result of the limitations of the circuit techniques used. Overall, the synchro output is an analog signal and as such theoretically contains all the information necessary to provide a digital output of unlimited resolution. Some of the errors under this heading, such as quantising, noise, sampling errors etc. have already been discussed in Chapter 5. Other errors could also occur such as calibration errors, the difference between the absolute value of θ and the actual value of the output, and additional errors caused by gain variations, settling-time and aperture errors.

Staleness is the error caused by the output representing the angle θ only at the instant at which it was sampled, and any subsequent changes are not reflected in the output reading until the next sample is taken.

Incorrect loading of the synchro and converters can also cause conversion errors and these are avoided by ensuring that manufacturers loading instructions are fulfilled.

Monotonicity is the characteristic of a converter which ensures that an 'increase' in the analog input signals will never cause a 'decrease' in the digital output. Harmonic-oscillator converters and pulse count techniques have perfect monotonicity whereas successive approximation converters can cause errors of this type unless specially protected.

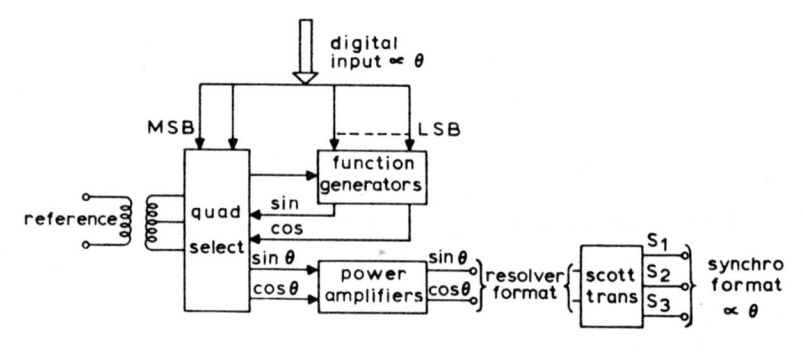

Fig. 6.23 Digital/synchro converter

Synchro/resolver conversion   161

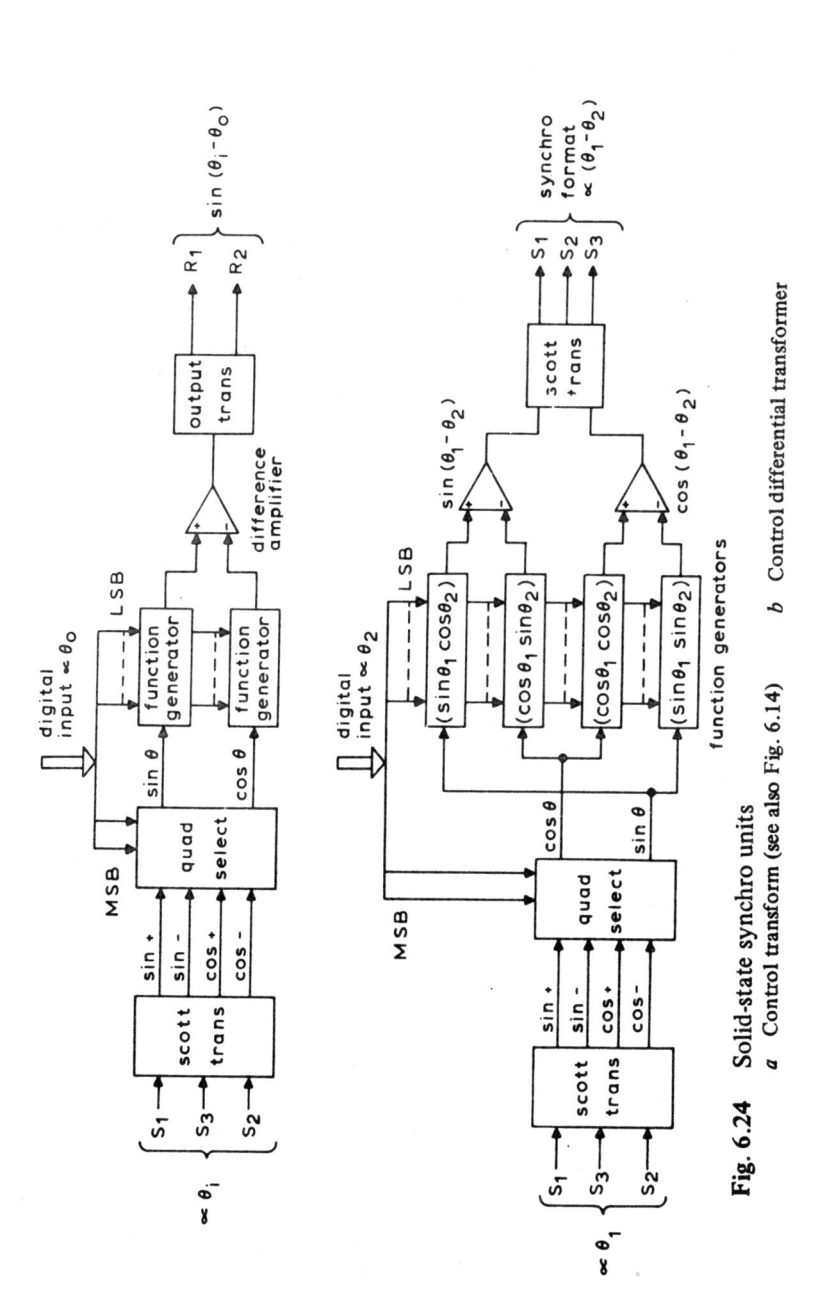

Fig. 6.24 Solid-state synchro units
a Control transform (see also Fig. 6.14)   *b* Control differential transformer

Digital-to-synchro conversion

The function generators referred to earlier, are, in fact, digital-to-synchro converters in that they can convert an incoming digital signal to an analog output of synchro or resolver format. These are available as separate units together with a variety of other digital/analog units which enable complex systems to be assembled operating primarily as digital signals.

The standard digital-to-synchro converter is shown in Fig. 6.23 which would form a self-contained unit providing outputs of synchro or resolver format referred to a common reference supply.

In using similar techniques it is possible to produce a solid-state control transformer having a standard analog synchro input θ and a digital input $\alpha\phi$, the output is an analog signal representing $\sin(\theta-\phi)$, corresponding to a conventional control transformer. Similarly, electronic control differential units are available. The block diagrams for these units are shown in Fig. 6.24a and 6.24b, respectively.

Chapter 7

Other techniques

In an ideal situation a digital transducer provides a continuous digitally-coded word representing the parameter being measured. The transducer has a high resolution and almost infinite sampling rate. Further, in the ideal transducer the information is not lost by temporary interruption of the power supply or the signal line. Such performance can only be achieved by absolute direct digital transducer, such as the optical encoder discussed in Chapter 2, and its linear equivalent. At the present time there are not many alternatives. This chapter describes some of the methods that have been adopted with some success. Other techniques that have been tried, but with less success, include photographic, persistent internal polarisation of phosphors, magneto-optical, thermoplastic, Curie point and colour centre recording. These are not discussed any further as there does not appear at this time, to have been any development to a stage where application to a practical digital transducer seems possible.

7.1 Digital position sensors

These devices, primarily used as fluid level detectors, sense the presence of a fluid by the interference it causes between small transmitting and pick-up transducers. The principle is illustrated in Fig. 7.1 in which the level of the column of fluid is sensed by the presence or absence of a signal in the pick-up units. Adaptations, as shown in Fig. 7.2, have been made which allow the principle to be adapted for measurement of pressure and temperature.

These units are more generally used as indicators or warning devices

164   Other techniques

and also to initiate control action through a direct digital link to a computer. For example, the absence of logic '1' on A in Fig. 7.1 could be used to indicate a dangerously low level. Logic '1' on C could indicate high, and logic '1' on D indicate a dangerously high level.

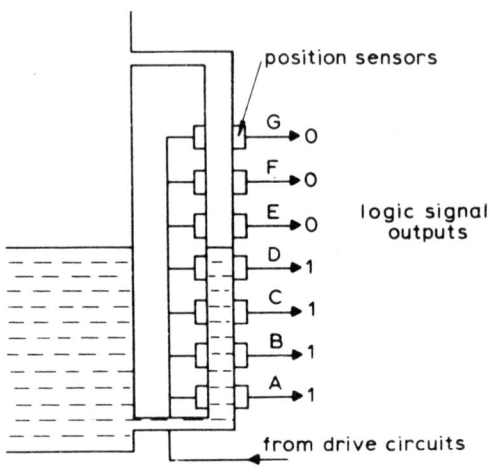

Fig. 7.1 Digital level indicator

When used as a direct digital transducer, the physical size of the sensors limits their application, and the resolution is only equal to the number of elements used. The outputs of the sensors can be combined to form a straight binary or Gray code digital output. For example, a device having 7 sensors can be converted to a three bit number by conventional logic; e.g.

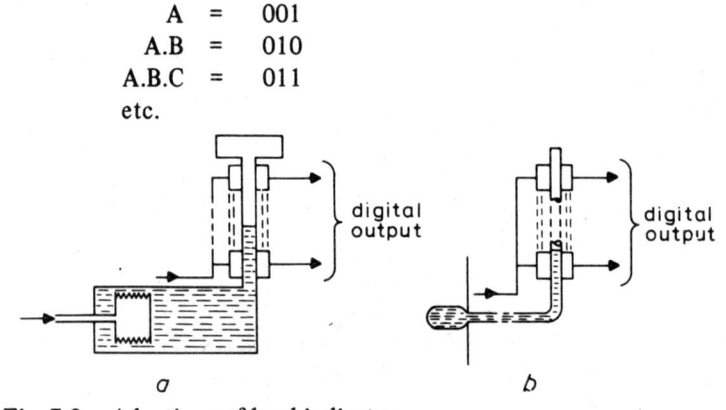

$$A = 001$$
$$A.B = 010$$
$$A.B.C = 011$$
etc.

Fig. 7.2 Adaptions of level indicator
    a   Digital pressure indicator
    b   Digital temperature indicator

A system of this type has been used in the computer-based data logging system installed in the nuclear powered merchant ship *N.S. Savannah* (Gertz and Leavitt, 1964). This system uses ultrasonic transmitters and sensors operating at frequencies in the order of 3 MHz. The transmitters and receivers are interchangeable and use a lead-zirconate-titanate piezoelectric material. The signal pulses are passed through about 6 mm, or less, of the fluid, which is sufficient to detect the presence or absence of the fluid.

Each of the transmitters are given a short burst of energy, at 3 MHz, lasting approximately 2 μs and this is repeated at a frequency of 60 Hz. This effectively represents a sampling rate, of the complete transducer system, of 60 samples/s.

Inductance sensors

A similar apparatus is described by Williams (1967) which has been used to measure the sea-water/oil interface in the waste-oil collecting tanks in submarines. The conductance between two sensing coils will change according to the fluid, therefore the presence of sea water or oil can be determined. Seven sensors are located to cover the complete depth of the tank; an additional unit situated at the bottom of the tank, which should be in sea water, is used for reference.

The primaries of all sensors are wired in series and energised by a continuous signal of 100 kHz (Fig. 7.3). Although each secondary can generate a separate 'bit', (logic '0' of logic '1') depending on whether

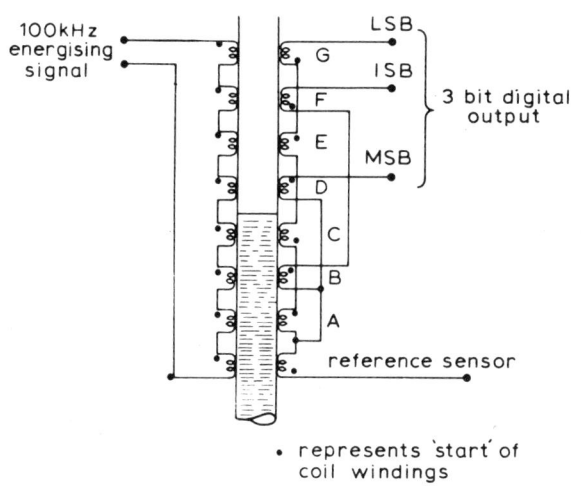

Fig. 7.3 Conductance level indicator

the coupling is oil or water, the sensors in this application are wired to produce a 3 bit Gray code directly. The secondaries are arranged in a series connection and in phase opposition as shown. The phases of the output are indicated by the dots on the coil ends. The magnitude of the output of the reference sensor has half the ouput of the other sensors. Relatively positive output indicates a logic '1' and a negative logic '0'. The half output of the reference sensor provides the positive and negative sensing for the other bits of the transducer output. Consider the output of the reference sensor as - 0·5 units when the fluid has passed that level, and that all the other sensors will produce 1 unit when the fluid has reached that level. At any level of the fluid, the MSB of the output is made up of the outputs from the reference sensor and sensor D only. The ISB (intermediate significant bit) is made up from the outputs of the reference sensor and sensors B and F, and the LSB is made up from the output of the reference sensor and sensors A, C, E and G. As an example, consider a level between C and D as shown in Fig. 7.3. There will be outputs from sensors A, B and C as well as the reference sensor:

Outputs:

Reference = $-0·5$; A = $+1$; B = $+1$; C = -1; D = E = F = G = 0.
MSB = Reference + D = $-0·5 + 0 = -0·5$ Logic 0
ISB = Reference + B + F = $-0·5 + 1 + 0 = +0·5$ Logic 1
LSB = Reference + A + C + E + G = $-0·5 + 1 - 1 + 0 = -0·5$
Logic 0
Logic output = 010 (Gray code) = 3_{10} = 3rd detectable level.

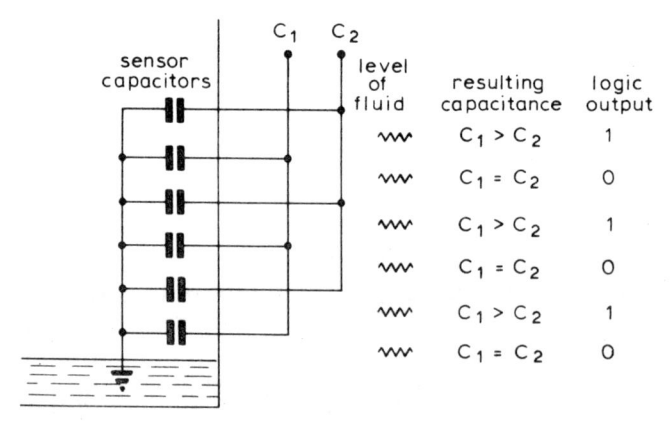

Fig. 7.4 Capacitance level indicator

Similar analysis will show the output progresses in a 3-bit Gray code format.

The electronic circuitry consists of an oscillator, phase comparators and the usual squaring and logic network. The output is continuous in the sense that the information is not affected by any temporary interruptions. The number of sensors can be increased according to the resolution required.

Capacitance sensors

Bowden and Johnston (1968) describe a similar method, but by using capacitance sensors the level measurement can be extended to powders and free-flowing granular materials. Detection is by the change of capacitance when the dielectric between the electrodes is changed by the presence of the fluid or other material.

As in the previous transducers, the connections allow the output to be made available as a Gray coded signal. For this purpose it is necessary to connect the sensors in a capacitance unbalance bridge. Fig. 7.4 shows the connection of six sensors which can be used in a bridge circuit to provide a single digit output which changes '0' to '1' to '0'

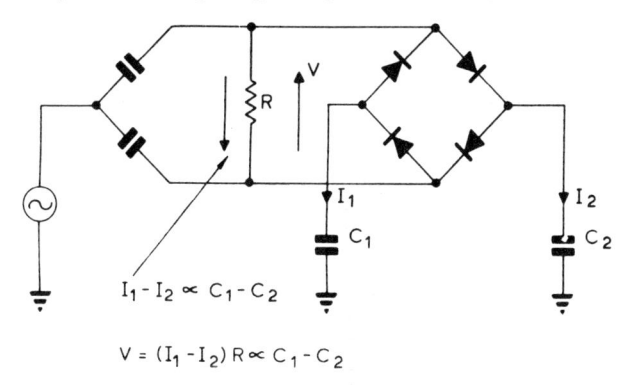

$$I_1 - I_2 \propto C_1 - C_2$$

$$V = (I_1 - I_2) R \propto C_1 - C_2$$

Fig. 7.5 Capacitance bridge

etc. as the level progresses in the six separate measurable levels. A circuit such as that shown in Fig. 7.5 can be used to compare the resultant capacitance C_1 and C_2. The voltage V developed across the resistor R is used as the logical output.

The resolution depends on the number of capacitors that can be connected in parallel, and this is generally limited. This is due to the overall unbalance of $(C_1 - C_2)$ which drifts away from zero line, and hence from the threshold value which defines logic '1' or logic '0'. This is due to the number of capacitors in the parallel circuits which change with level. The drift due to this cumulative effect is shown in Fig. 7.6. The maximum number of capacitors that can be used depends on the sensitivity of the bridge circuit and the type of electrodes used. These details of the system also depend on the fluid on which the measurements are being made.

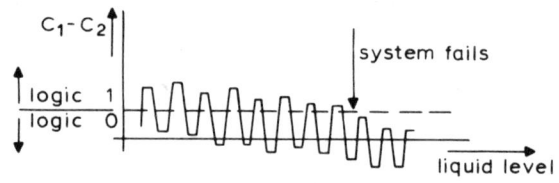

Fig. 7.6 Effect of cumulative tolerances

The connections required to provide a 3 'bit' Gray code from seven capacitor sensors is shown in Fig. 7.7. The electronic components not shown, include a signal generator, a bridge network for each 'bit', squaring circuits etc. The environmental conditions appropriate to the circumstances can usually be catered for by a suitable choice of electrodes. For example, flat plastic coated electrodes can be used in corrosive conditions. Similarly, special designs can be prepared to work under other special conditions.

7.2 Force balance feedback transducer

The transducers described above are essentially open-loop devices,

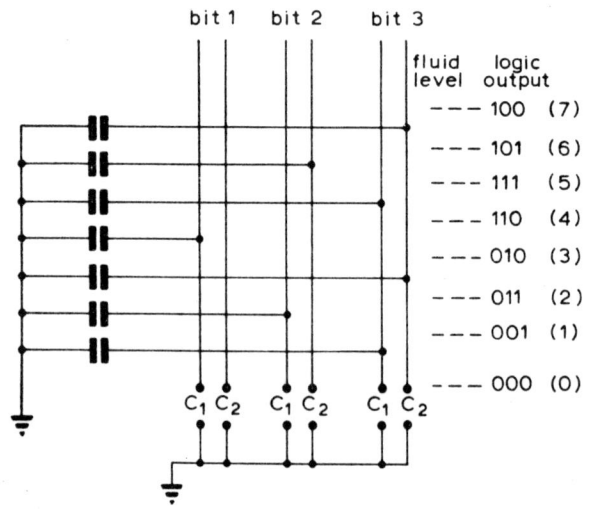

Fig. 7.7 Digital capacitance level transducer

which is the usual technique adopted in any transducer. It is generally accepted that the introduction of some form of feedback would allow an increase in the accuracy and, sometimes, the resolution of a transducer system. In feedback systems the essential feature is that the measured variable initially appears as an analog signal, which is converted to a digital output, and also it includes a D/A converter which converts the digital output back to an analog signal which can be compared with the original analog signal. A successful development of this idea* (Serra, 1966; Sherwood, 1969) uses the force balanced principle to convert an input pressure to a parallel digital signal. Since there already exist a number of methods of measuring temperature, flowrate etc. as a pressure difference, this technique can be used for the measurement of a variety of different variables.

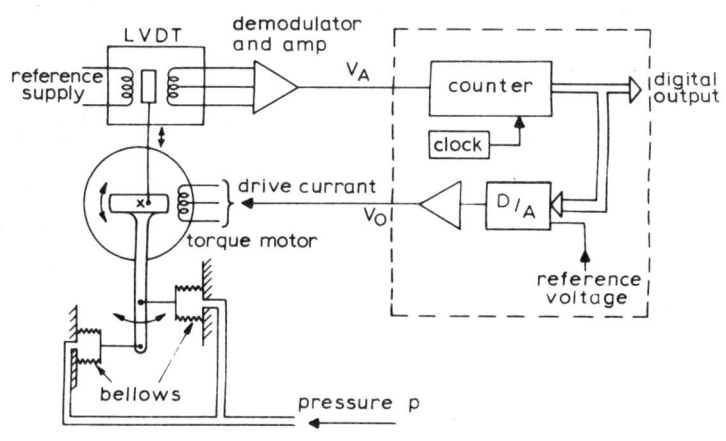

Fig. 7.8 Force balance digital transducer

The principle of the unit is shown in Fig. 7.8. The left-hand side of the diagram shows the electromechanical force-balance system and the right-hand side the digital part which behaves in exactly the same manner as the feedback A/D converter described in Chapter 5. An analog voltage V_A is used to increase or reduce the count in a counter. The counter changes until V_A is reduced to zero. The count at any instant is converted by a D/A unit to an analog voltage V_0 which is used by the electromagnetic system to reduce V_A to zero.

The electromagnetic system is therefore acting as the voltage comparator of the conventional feedback A/D converter. The comparison between V_A and V_0 is effectively done by the bellows assembly, the torque motor and the linear variable differential transformer (LVDT).

*by the Conrac Corporation

An input pressure P is converted to a force by the two sets of bellows. This force generates a torque on a shaft that carries the moving part of a torque motor. The shaft rotation also causes displacement of the core of the LVDT. Displacement of this core from the central position disturbs the equality of the flux in the secondary coils and the output of these coils is sensed and compared to the phase of the primary reference supply to generate the voltage V_A. This voltage has a magnitude proportional to the displacement of the core of the LVDT (linear variable differential transformer) and a sign according to an upward or downward displacement.

The A/D converter behaves in the usual manner to produce a voltage V_0 which is used to drive a current through the torque motor to produce a torque opposing that developed by the bellows. Equilibrium is reached when these torques are equal and the LVDT is in the zero position, in which case V_A is zero. When this occurs, V_0, and therefore the digital output, is a direct measure of the torque and hence the pressure P.

In the actual transducer, the torque motor is similar to a normal moving-coil type instrument and is quite small. The whole of the electromechanical assembly, the bellows, LVDT, and torque motor are assembled into a unit only 6·5 mm long. The electronic circuits are complex and involved for a single transducer, which tends to increase the overall cost of the instrument. Sharing some of the electronic circuits with other similar transducers helps to keep costs down, but the system then has only marginal advantages over more conventional analog transducers with conventional A/D converters.

Performance for a range of three transducers measuring pressures up to 70 atmospheres is quoted as having an accuracy of 0·05 per cent with a resolution of 1 part in 8000. Temperature coefficient of 0·001 per cent is quoted, and this is achieved by special electrical compensation, although in the uncompensated state the temperature coefficient might be sufficiently satisfactory under a number of normal industrial conditions. The instrument takes approximately one second to give a zero to full-scale reading, which indicates a low frequency response. This is governed by the necessity of maintaining adequate damping in the electromechanical system and also by the clock pulse rate.

7.3 Magnetic transducers

Shaft encoders using magnetically sensitive discs, as described in Chapter 2, have been used for a number of years, and similar devices to measure linear displacement have been proposed. The only serious attempts at a high resolution linear magnetic digital transducer seem to have been made in the USA under military development contracts (Bossard, 1962; Schuster, 1963), both dated 1963. Considerable developments have taken place in digital recording techniques in recent years, primarily due to the requirement of the computer industry, and this has led to the widespread use of high-density recording. Tape, drum and disc stores are now functioning with densities of well over 40 bits/mm, with over 25 tracks on a 6 mm wide tape. Densities of 100 bits/mm are easy to achieve, and it has been suggested that the theoretical limit is in the order of 8000 bits/mm although this is unlikely to be achieved. Hence, there is no real problem in recording techniques to achieve a high linear resolution with a high degree of accuracy as regards the position of the tape or disc relative to the recording heads. The digital information can be recorded on the magnetic surface (disc or tape) during calibration to represent the physical displacement, and any desired code may be used. The technique is, therefore, not subject to any calibration errors. These advantages do not seem to have been exploited except in the experimental device referred to earlier.

Magnetic matrix transducer

This transducer (Bossard, 1962) uses a matrix of wire rods which are wrapped with wires and 'read' in a manner similar to the way in which magnetic core stores are used in digital computers. An armature moves over the matrix influencing parts of the matrix to register either logic '1' or logic '0' depending on the armature position. The displacement of the armature is measured by 'reading' the output of the matrix. The sense windings of the matrix are cross connected to provide the desired codes. The magnetic energy is provided by an alternating magnetic field generated in the armature.

The accuracy of the apparatus depends on the precise construction of the matrix assembly. It also follows that the matrix becomes more difficult to assemble as the number of 'bits' increases. The connection of the read windings also becomes more critical and it becomes more difficult to differentiate between logic '1' and logic '0'. In an actual model a resolution of 0·1 mm of armature travel was the best that could be achieved, the transducer having a total travel of about 12 mm, thus providing a seven bit digital readout. In this model the rods were 0·22 mm in diameter, wrapped with wire 0·08 mm diameter, and the armature energised with pulses of 3 μs at 30 kHz. The signal levels

172 Other techniques

were very small and required amplifying and shaping. Attempts with smaller rods and wires did not prove practical although it is considered that a manufacturing technique could be found.

Accuracy and repeatability of this design depends largely on the mechanical arrangement used. Only one example is quoted in detail and this was an altimeter which is effectively a pressure transducer (Fig. 7.9) using an aneroid pressure capsule to produce a physical displacement of the armature. The resolution, accuracy and repeatability were not high enough to make the instrument sufficiently promising unless further work on the matrix assembly to improve these factors is done. The overall accuracy was not better than 2% and a high temperature coefficient was observed.

The report ends with a recommendation that further investigation should be conducted into this and other types of magnetic transducers. In view of more recent developments in core stores, it seems likely that renewed interest in the application to transducers could lead to considerable developments.

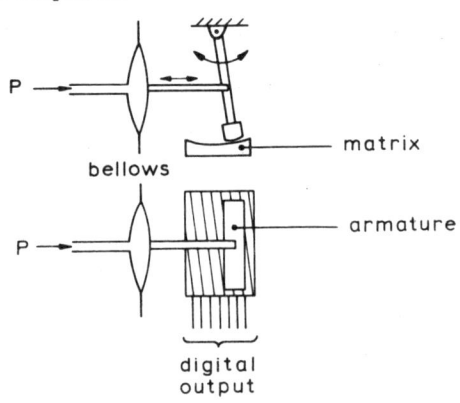

Fig. 7.9 Magnetic transducer

Magnetic recording transducer

The transducer* described by Schuster (1963) depends on small magnetically sensitive surfaces assembled in the manner shown in Fig. 7.10. Ten such surfaces were in fact used, fixed to a carriage secured to the actuating shaft which is moved by the displacement to be measured. A magnetic recording/read head rests on each surface, with a pressure pad to ensure constant contact.

The digital data is recorded as one bit on each sensitive surface

*developed by Radiation Inc.

through the specially designed heads and a unique electronic arrangement. A different digital word can be recorded for each unit of physical displacement during calibration. As a transducer, the digital information is reproduced by the head being used in a reading mode.

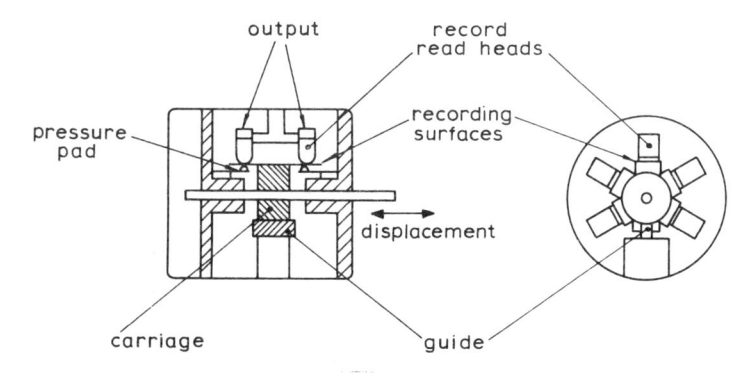

Fig. 7.10 Magnetic transducer with recording replay

With the actual model constructed, the resolution was 0·005 mm. The overall range depends on the number of recording surfaces used. The temperature coefficient was high on this model but with suitable redesign, acceptable values could be reached. The mechanical arrangement does not suggest that the transducer would be satisfactory under conditions involving vibration or shock loads. Quite clearly, friction to motion of the input displacement and wear, due to rubbing between the head and magnetic surface, would be a serious problem in a commercial transducer.

The principle, however, has been demonstrated to be satisfactory and it can be expected that further developments using other types of heads etc. will lead to an acceptable high accuracy transducer.

7.4 Radiation transducers

Transducers using a radioactive source have only recently been introduced and the account given by RAO (1970) summarised the position at the University of Strathclyde where a particular application of this type of unit has been under development.

The simplest form of transducer of this type is that shown in Fig. 7.11 and consists of a radioactive source S, a variable aperture and a detector. Variation of the position or area of the aperture produces a change in the radiation reaching the detector. Radiation striking the detector produces a random sequence of pulses which can be counted over a prescribed time interval. Providing the time is sufficiently long, the total is the output of the transducer and represents the average radiation reaching the detector. The relationship of output to system parameters is given by

$$P_0 = ksD/d^2 \qquad (7.1)$$

where S = source emission
 P_0 = 'output' of detector, i.e. count over a given sampling period
 k = constant
 D and d are as shown on Fig. 7.11 and represent alternative transducer 'inputs'.

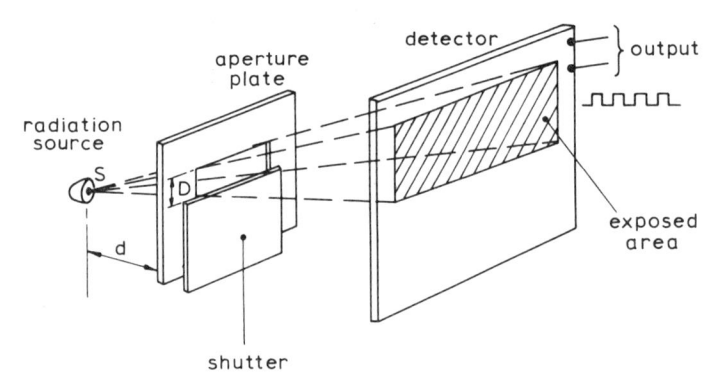

Fig. 7.11 Radiation transducer

The advantages claimed for this type of transducer are as follows:

(i) independent of environment as the emission rate at the source is independent of pressure, temperature, etc.;
(ii) rugged, compact and lightweight (but not miniature);
(iii) high degree of calibration accuracy;
(iv) infinite resolution as with frequency domain transducers, providing sufficient counting time is allowed.

The radiation source must be contained in a suitable shield and consideration given to its half life since decreasing radiation due to natural decay will affect output. Periodic recalibration appropriate to the source half life will overcome this difficulty. The transducer input is a mechanical displacement which will cause either a change in the distance d or the effective aperture size D. The primary source of the displacement could be derived from bellows which thus extend the range of the transducer to the measurement of pressure, temperature, fluid flow etc.

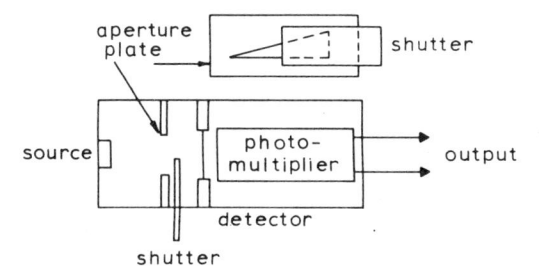

Fig. 7.12 Construction of radiation transducer

A particular advantage is the comparative ease with which nonlinear functions or laws can be accommodated. It is only necessary to shape the aperture profile such that the open area follows the required relationship to the input displacement. Some constructional details of an actual transducer are shown in Fig. 7.12. The shutter moves over a shaped aperture interrupting the radiation to a zinc-sulphide detector which reacts to a photomultiplier tube to increase the output intensity. In some transducers working on this principle, it has been found that counting times of the order of one second are necessary to achieve 0·1 per cent accuracy. This time will govern the maximum sampling rate, but it will depend on the intensity of the radiation source and the distance between source and detector. Good results can be achieved with the system containing air at ordinary pressure, but some advantage is gained by evacuating the system so that the output is decided entirely by the geometry and is not influenced by collisions of the radiation particles with gas molecules.

Radioactivity as a source of pulses has also been used for flow measurement. The system suggested consists essentially of a neutron source of radiation located upstream of a detector. The source will cause some of the particles of the fluid to radiate detectable particles some of which will be reabsorbed. However, some will still be active when they pass the detector, giving pulses that can be counted over a period of time. For a given fluid, pipe size and distance between source and detector, the count over a given period will be a measure of the flow rate.

This method of flow measurement could be of particular value for slurries and semi-liquid materials, but obviously suffers from the same drawback as the previous radiation transducer, i.e. resolution and accuracy is dependent on the time period of the count.

7.5 Vortex transducers

These devices are applicable basically to fluid flow and generate an output whose frequency, and therefore periodic time, is a function of the flow rate. One of these principles, referred to by Sawyer *et al.* (1975), depends on the generation of vortices shed by a shaped body in a stream of fluid. The vortex pattern depends on the body shape and properties of the fluid, while the frequency of the vortices depends on the flow rate for that particular fluid (Fig. 7.13a).

The frequency can be detected as a pressure variation, using a piezoelectric pressure transducer or by self-heating thermistor sensors. The latter consists of a heated thermistor whose change in resistance depends on the changing cooling rate induced by the vortex. The changes in pressure on the piezoelectric crystal, or changes in resis-

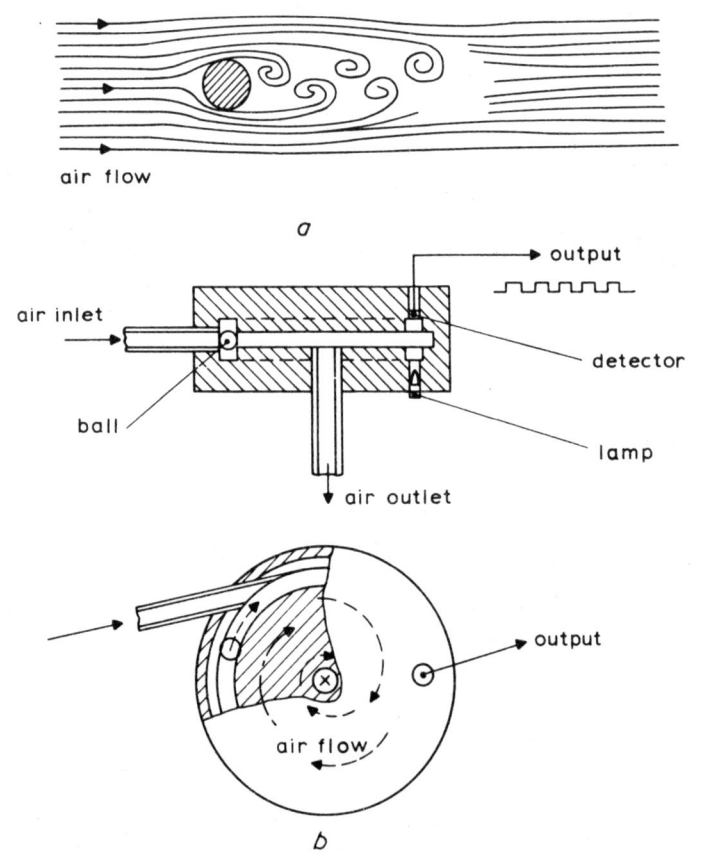

Fig. 7.13 Vortex transducers

tance of the thermistor, can be used to measure the periodic time of the vortex frequency or to accumulate a count. In either case, the principles described in Chapter 3 can be used to give a measure of the vortex frequency and hence the fluid flow rate.

An alternative method which has a particular application to low flow rates is to use a vortex chamber in which a vortex is induced in a circular chamber, as illustrated in Fig. 7.13b. The fluid is necessarily caused to rotate in a vortex pattern and this induces the ball to move around its circular track. Each time the ball interrupts the light cell unit a pulse is generated, and these pulses can, as before, be counted or the periodic time measured to give a measure of the flow rate.

7.6 References

BOSSARD, C.L. (1962):'Digital transducer research program 5935-M, final report', Bureau of Naval Weapons Report NADC-RP-L6291 AD-295 673

BOWDEN, K.R.R., and JOHNSTON, J.S. (1968): 'Novel capacitance level-sensing transducers' *in* 'Industrial measurement techniques for on-line computers'. IEE Conf. Publ. 43, pp. 22-29

GERTZ, D., and LEAVITT, L. (1964): 'Standardising instruments– digital systems', *Electronics*, 37, 65-70

RAO, D.N. (1970): 'Digital transducers using a radioactive source', IEE Electronics Division discussion.

SAWYER, P.E., LOCKETT, A.D., and THOMPSON, J.W. (1975): 'Some trends in computer-based laboratory automation' *in* 'Trends in on-line computer control systems'. IEE Conf. Publ. 127, pp. 246-252

SCHUSTER, R.R. (1963): 'Experimental study of application of magnetic recording to digital transducers', Radiation Incorporated Technical Documentary Report ASD-TDR-63-381

SERRA, G.F. (1966): 'Force-balance principle raises accuracy of digital pressure transducer', *Instrum. Soc. Am.*, 13, pp. 51-54

SHERWOOD, W.M. (1969): 'The search for a true digital transducer', *Control Eng.*, pp. 95-98

Appendix 1 Binary codes

Straight binary

In a natural or straight binary code the weighting of each bit is derived as follows:

$$\text{MSB} \quad \underbrace{2^7 \quad 2^6 \quad 2^5 \quad 2^4 \quad 2^3 \quad 2^2 \quad 2^1 \quad 2^0}_{} \quad \text{LSB}$$

with the least significant bit (LSB) traditionally on the right and the most significant bit on the left. Table A.1 gives the straight binary numbers up to 31_{10} which requires five bits.

For convenience, a binary coded word is often referred to in its octal coded version in which the bits are grouped into threes, and the word represented in its decimal form. As three bits can only represent up to the decimal number 7, then 8 and 9 never appear.

For example, the decimal number 2807_{10} is represented as 5367 in octal. This is derived from the straight binary as follows:

$$2807_{10} \;=\; \underbrace{101}_{5_{10}} \; \underbrace{011}_{3_{10}} \; \underbrace{110}_{6_{10}} \; \underbrace{111}_{7_{10}}$$

$$= \; 5367_8 \text{ in 'octal'}$$

Binary coded decimal (BCD)

A binary coded decimal code is a convenient way of grouping a digital code when it is finally required to display the number in its decimal form. Four bits will be required for each digit. Thus, a 12 bit number can only represent the maximum decimal equivalent of 999_{10}. For example:

$$\underbrace{1001}_{9_{10}} \underbrace{0111}_{7_{10}} \underbrace{0001}_{1_{10}}$$

$$= 971_{10}$$

Table A.1 Comparison of decimal, binary and Gray codes

DECIMAL	BINARY	GRAY
31	1 1 1 1 1	1 0 0 0 0
30	1 1 1 1 0	1 0 0 0 1
29	1 1 1 0 1	1 0 0 1 1
28	1 1 1 0 0	1 0 0 1 0
27	1 1 0 1 1	1 0 1 1 0
26	1 1 0 1 0	1 0 1 1 1
25	1 1 0 0 1	1 0 1 0 1
24	1 1 0 0 0	1 0 1 0 0
23	1 0 1 1 1	1 1 1 0 0
22	1 0 1 1 0	1 1 1 0 1
21	1 0 1 0 1	1 1 1 1 1
20	1 0 1 0 0	1 1 1 1 0
19	1 0 0 1 1	1 1 0 1 0
18	1 0 0 1 0	1 1 0 1 1
17	1 0 0 0 1	1 1 0 0 1
16	1 0 0 0 0	1 1 0 0 0
15	0 1 1 1 1	0 1 0 0 0
14	0 1 1 1 0	0 1 0 0 1
13	0 1 1 0 1	0 1 0 1 1
12	0 1 1 0 0	0 1 0 1 0
11	0 1 0 1 1	0 1 1 1 0
10	0 1 0 1 0	0 1 1 1 1
09	0 1 0 0 1	0 1 1 0 1
08	0 1 0 0 0	0 1 1 0 0
07	0 0 1 1 1	0 0 1 0 0
06	0 0 1 1 0	0 0 1 0 1
05	0 0 1 0 1	0 0 1 1 1
04	0 0 1 0 0	0 0 1 1 0
03	0 0 0 1 1	0 0 0 1 0
02	0 0 0 1 0	0 0 0 1 1
01	0 0 0 0 1	0 0 0 0 1
00	0 0 0 0 0	0 0 0 0 0

180   Appendix 1

Each group of four bits is not used to its fullest extent as only ten combinations are required. There are therefore six redundant combinations and the system is therefore inefficient. In the example given, the twelve bit natural binary code can represent up to 4095_{10} whereas in the BCD word the maximum number is 999_{10}. In the BCD number 971_{10} above each decimal digit has a binary word following the natural binary code, the bits represent 1, 2, 4 and 8. As only 10 separate combinations of the four bits are needed many other weightings are possible, but the 1, 2, 4 and 8 is by far the most popular in current use.

Gray codes

These codes have developed from the original Gray code devised by Frank Gray. The Gray code is a reflective binary code, i.e. in changing from one value to the next increment only one bit is changed at a time. In the natural binary many bits may have to change for a single increment. For example, in a 4 bit number in natural binary code, when changing from 6_{10} to 7_{10} only the LSB changes;

i.e. 6_{10} = 0110
     7_{10} = 0111

But in change from 7_{10} to 8_{10} all four bits have to change;

i.e. 7_{10} = 0111
     8_{10} = 1000

An error of only one bit in a large digital number can cause large errors in the decimal reconversion. Reflective codes reduce these errors, particularly in the case of transducers where an increment in the measured variable should give a change in only one digit.

Column 3 of Table A.1 shows a Gray code for the decimal numbers 0_{10} to 31_{10}. It is evident that the MSB is identical to the natural binary but all other numbers are different.

The conversion between natural binary and Gray code can be made as follows. First ensure that each coded number contains the same number of bits, i.e. all leading zeros must be present. Proceed as follows:

(a)   Natural binary to Gray code

(i)   MSB is unchanged;

(ii) for each other bit in the natural binary, the Gray code bit is the same if the digit to the left in the natural binary is a zero;
(iii) if the bit to the left is one, then the bit is changed.

(b) Gray code to natural binary

(i) MSB is unchanged;
(ii) for each other bit in the Gray code, the natural binary is the same if the number of ones to the left is even;
(iii) if the number of ones to the left is odd then the bit is changed.

Examples:

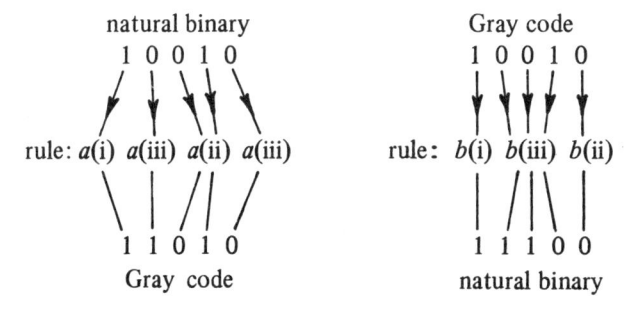

```
        natural binary                    Gray code
          1 0 0 1 0                       1 0 0 1 0
rule:  a(i) a(iii) a(ii) a(iii)      rule:  b(i) b(iii) b(ii)
          1 1 0 1 0                       1 1 1 0 0
           Gray code                      natural binary
```

Excess Gray codes

In some digital transducers it is found necessary to have a BCD type of output with each decimal digit represented by a four bit Gray code. Each four bit code must be cyclic, that is only one bit changing for each successive decimal digit, and this will now include the change from 9_{10} to 0_{10}. The usual Gray code of Table A.1 will not allow this as:

$$0_{10} = 0000$$
$$1_{10} = 0001$$
$$\vdots \quad \vdots$$
$$5_{10} = 0111$$
$$\vdots \quad \vdots$$
$$9_{10} = 1101$$
$$0_{10} = 0000$$

i.e. three of the four bits will need to change. This can be overcome by using the excess 3 code which is obtained by adding three to each decimal digit and then using the Gray code equivalent.

Appendix 1

Decimal number	Excess 3 decimal	Gray code (excess 3)
0	3	0010
1	4	0110
2	5	0111
3	6	0101
4	7	0100
5	8	1100
6	9	1101
7	10	1111
8	11	1110
9	12	1010
0	3	0010

Thus, cycling 0_{10} through to 9_{10} and again to 0_{10} etc. only one bit is changed at each decimal digit. A slight advantage also occurs that zeros on all four signal lines can only indicate a fault as, to be correct, the Gray code must always contain one or more '1's.

This code is perfectly satisfactory for single digit decimal numbers. If, however, a decimal number has two or more digits, each digit will require four bits. When the overall decimal number changes through each decade, two decimal digits will change and on the overall excess 3 code more than one bit will change, since there is a 1-bit change for a change in each decimal digit. A limited number of such ambiguities arise, for example in changing

from: 179_{10} – Gray excess 3 code – 0110 1111 1010 (8 x 1s)

to : 180_{10} – Gray excess 3 code – 0110 1110 0010 (6 x 1s)

Thus, there are two bit changes for a single increment in the decimal number. This type of ambiguity can be overcome by adapting a 'decimal unit distance' code. This method converts the digits of the decimal number, before converting it to a binary number, using the Gray excess 3 code. Each digit of the decimal number is examined and if its more significant digit (the digit on its left) is even, the nines complement of the digit is written down. Leading zeros must be included; e.g. if the decimal number can have 3 digits then zeros must be inserted to give three digits in total. (The nines complement is the difference between the value of the digit and 9.)

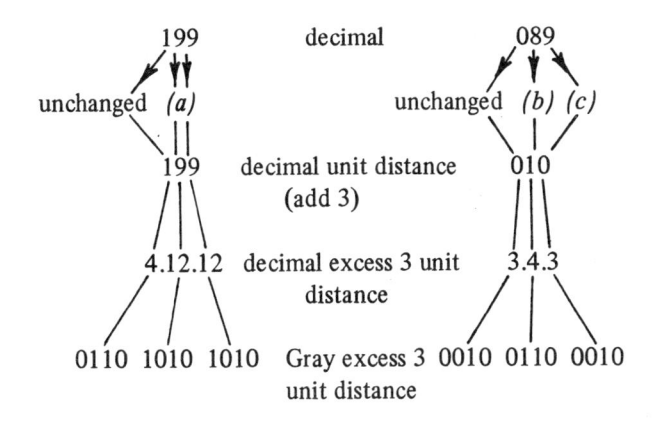

```
        199           decimal              089
       /↑↑                                /↓↓
unchanged (a)                    unchanged (b) (c)
       \ ||                               \ | /
        199    decimal unit distance       010
       /|\           (add 3)              /|\
      / | \                              / | \
    4.12.12   decimal excess 3 unit    3.4.3
     /|  |\          distance          /| |\
    / |  | \                          / | | \
 0110 1010 1010  Gray excess 3   0010 0110 0010
                 unit distance
```

(a) significant digit is odd;
(b) significant digit is even, hence 'nines' complement = 9-8 = 1;
(c) significant digit is even, hence 'nines' complement = 9-9 = 0.

Using this code some other decimal numbers may be coded:

Decimal	Gray excess 3 unit distance		
999	1010	1010	1010
000	0010	1010	1010
010	0010	1110	0010
100	0110	0010	1010
199	0110	1010	1010
200	0111	1010	1010

It will be found that for all decimal numbers 000_{10} to 999_{10} and back to 000_{10} there will be a change in only one bit for each unit change in the decimal number.

Note that these codes refer only to a three decade decimal number. The code would not be correct for the decimal number 200_{10} in a four decimal decade system as this should be written as 0200_{10} before coding.

Other codes have been devised to overcome ambiguities that arise in special cases. There are techniques which assist in decoding these special codes and in the design of logic systems to translate the coded binary number into a form suitable for measurement or control purposes.

Appendix 2 Analog transducers

A transducer can be defined as a device which converts an input signal to an output signal of a different form. In this text this is limited to devices having an electrical output while the input is usually some physical property such as pressure, fluid flow, or temperature or some mechanical displacement which itself may be caused by a change in some physical property.

One of the important characteristics of a transducer is that it should have a negligible effect on the parameter being measured. This generally implies that it will impose no physical force or cause any change in the physical characteristics of the parameter. This requires that moving parts shall have small mass and that friction and other restraining forces shall be negligible.

In the majority of the transducers considered here, the output will be a voltage such that, for a linear relationship, the change in voltage

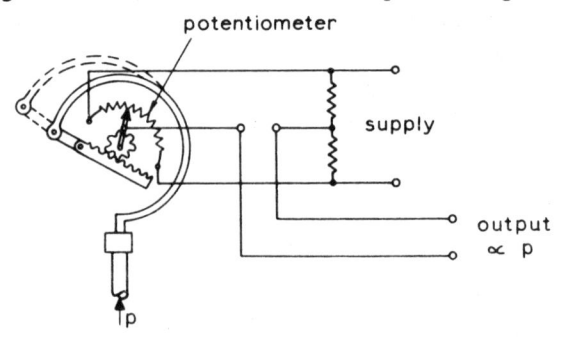

Fig. A.1 Pressure transducer using Bourdon tube and potentiometer

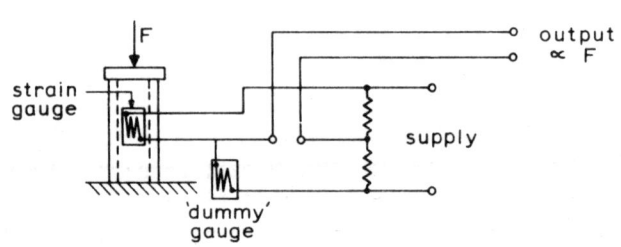

Fig. A.2 Force transducer using strain gauges

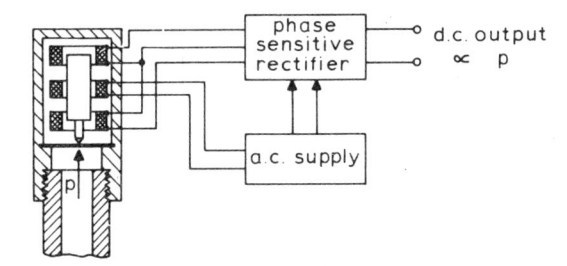

Fig. A.3 Pressure transducer using LVDT

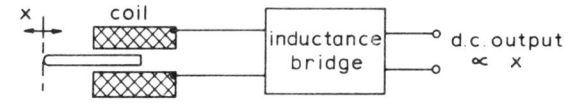

Fig. A.4 Displacement transducer using variable reluctance

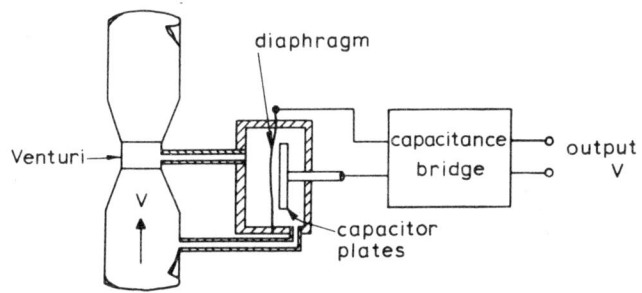

Fig. A.5 Venturimeter flowmeter using capacitance bridge

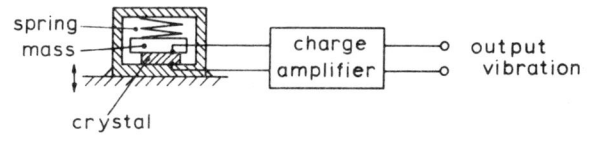

Fig. A.6 Vibration transducer using piezoelectric crystal

should be directly proportional to a change in the parameter being measured. Fig. A.2.1 illustrates some transducers which convert a physical characteristic to a voltage signal by some of the techniques described in the following paragraphs.

Resistance change

By far the most popular technique in use is the resistance change in the form of potentiometers or strain gauges.

For maintenance of linear performance, a potentiometer must be supplied from a constant current source and the wiper output connected to a high impedance instrument to minimise loading effects. The physical parameter to be measured must be converted to a displacement of the slider, usually through some mechanical linkage. The resistance itself can be wire wound, which generally has the longer life, or produced by depositing carbon on one of the resistance compounds. The force required to move the slider is often unacceptable in some transducers, although many physical parameters such as pressure, force, flow temperature etc. can be used to derive the necessary mechanical displacement. The resolution of a potentiometer depends on the mechanical arrangements and, in the case of wire wound potentiometers, by the wire size and the type of wiper used. A maximum resolution of 0·05 mm at the slider seems to be the practical limit at the moment, with an overall error of about 0·1%, depending, of course, on the stability of the supply voltage. The upper operating frequency is about 1 Hz primarily due to mechanical limitations.

Resistance strain gauges have been adapted to a wide range of instrumentation systems and many special applications have been devised. The strain gauge was originally a very fine wire or a layer of parallel wires connected in series. The wires, usually attached to a thin flexible insulated film, were glued to a surface. Any strain, i.e. change in dimension, in the surface caused a corresponding change in the resistance wire, thus giving a change in its resistance. This change in resistance can be detected by connecting the strain gauge into a bridge circuit, usually one arm of the bridge being a similar gauge attached to an unstrained surface. This second gauge provides automatic compensation for changes in temperature that would cause changes in resistance. By suitable arrangement of the bridge, temperature changes are effective on both gauges and cancel each other out.

Modern strain gauges are produced by printed circuit or other thin film techniques which enable considerable improvement to be achieved on the characteristics of the wire types. They can be made very small (2 mm if necessary) and also be produced conveniently as rosettes with 2, 3 or more gauges with different orientation. Normally, all types of gauges require special techniques in securing to the surface, and also sealing-over to prevent moisture affecting resistance changes. Making the necessary electrical connections also involves special skills and techniques.

The longitudinal axis of the gauge, in the simplest installation, is secured in the direction of the principle stress, and therefore the principle strain, in order to produce maximum change in resistance.

However, there are a variety of directions which are used to take advantage of special properties in some configurations.

Any parameter that can be made to produce a motion and/or a force, even if it is very small, can be used to develop a strain, usually in a metallic material, which can be measured by a strain-gauge system.

Resolution of transducers using strain-gauge techniques is usually limited by the measuring circuit and the stability of the voltage supply to the bridge. It also depends on the mechanism used to convert the parameter change to a strain-gauge resistance change. The output voltage across the bridge is usually of the order of 1 to 4 mV per volt of bridge supply, and some amplification is usually necessary. Errors of the order of ±1% are common up to operating frequencies of 10 kHz. The upper frequency limit is unlikely to go beyond 10 kHz as it is restricted by the limiting minimal size of the gauges themselves.

Capacitance and inductance changes

Capacitive transducers have been used specifically in the measurement of small displacements as might arise in measurement of mechanical vibration. In the simplest form they consist of two metal plates separated by a small air gap. The capacity is measured on an alternating current bridge and any change in the gap, or in the relative areas (by one plate moving across the other) will cause a change in capacity that can then be measured. The advantage of these devices is that there is negligible or no restraining force on the moving member, with little, if any, added inertia. For maintenance of accuracy, the characteristics of the dielectric, i.e. the air between the plates, must be constant and hence changes in relative humidity etc. can cause errors. The movement can either be measured by a self-balancing bridge or by using the capacitors to control the frequency of an oscillator, hence any change in capacitance will cause a change in frequency. This can have a direct application to digital systems.

Inductive transducers are similar in principle; a moving iron core causes a change in the inductance of a system of coils. This will also normally require an a.c. supply and some form of rectifying and/or phase sensing system similar to the capacitive transducers. Generally, there will be some small force opposing the displacement of the core in inductance transducers and this may be unacceptable in some cases.

The resolution of the capacitance and inductive transducers depends on the detecting system and accuracies in the order of ±0·5% can be expected.

A popular form of electromagnetic transducers are those commonly known as LDVTs - linear-differential-variable-transformers. They generally consist of primary and split secondary coils, physically separated on a nonmagnetic former. The ferromagnetic core provides the mutual coupling between the energised primary and the two secondary coils. Displacement of the core changes the induced voltages in the two secondary coils. The outputs of the secondary coils are connected in series so that when the core is in the central position the resultant output is nearly zero. Displacement from the centre causes the voltages in the secondaries to change, the resultant output voltage increasing, either in-phase or out-of-phase with the primary supply, according to the direction of the displacement from the central position. The accuracy and resolution depend on the stability of the primary energising supply and the phase sensitive network required to produce a d.c. output proportional to the displacement of the core. The transducer usually has only a small moving inertia and very little friction. Transducers are available with total displacement as small as 2 mm, the upper end of the range being about 50 mm.

For capacitive and inductive transducers and LVDTs, a frequency response up to about 2 kHz may be achieved depending on the frequency of the a.c. supply to the detecting system and the internal mass of the moving parts and other physical characteristics. Systems with a maximum error of 0·3% have been achieved.

'Synchros' are electromagnetic transducers used only for rotational motion. They require energisation by an alternating current and subsequent demodulation. These units are discussed in greater detail in Chapter 6.

Temperature measurement

There are several ways of deriving electrical outputs proportional to temperature. Normal liquid thermometers having a Bourdon tube indicator can be used with potentiometers, as in other Bourdon tube instruments, to provide an electrical output but these are not used a great deal. Resistance thermometers consist simply of a wirewound resistor whose change in temperature is sensed by a bridge network. Similar in use are thermistors that are specially manufactured resistors which provide a greater resistance change than ordinary wirewound resistance thermometers. These devices can be used to detect small changes in temperature, down to 0·001°C, but the upper temperature is generally limited to about 500°C for thermistors, and 700°C for resistance thermometers. The thermal capacity restricts the frequency response of these units to an upper limit of about 0·1 Hz.

Thermocouples can be used up to temperatures of 3000°C. They consist of a small junction of two dissimilar metals that develops a potential when the temperature is raised, usually relative to a similar junction maintained as some reference temperature. The two junctions are known, respectively, as the hot junction and the cold junction. The small potential is usually directly proportional to the temperature difference and an accuracy of 1% to 2% can usually be achieved. Hot junctions are enclosed in some protective sheath to insulate against mechanical damage, but this does often increase the time lag and the time constant can be one second or more. This naturally restricts the frequency response to less than 0·2 Hz.

Piezoelectric devices

Piezoelectric crystals, some occurring naturally and others manufactured, are processed by a special polarising treatment and selective cutting to produce a voltage across their opposite faces under the action of a force. For the small crystal used in transducers, the output voltages are usually measured in millivolts. The detecting network usually requires a charge amplifier or similar device due to the impedance characteristics of the crystals.

They have been used for a variety of transducers for measuring force and hence pressure, but due to difficulties in detection circuits at low frequencies (i.e. less than 1 Hz), their most popular use is in dynamic measurement where their high frequency response (up to 1 MHz in special cases) and small mass, makes them ideal for measurement of mechanical vibration. Accuracies of ±1% are normally achieved.

Semiconductor strain gauges, can also be classified as piezoelectric devices. These are crystals whose structure and electrical resistance change under the action of a force. When suitably cut crystals are used in this way they provide a strain gauge having an effective sensitivity 50 to 100 times that of normal resistance strain gauges. This enormous advantage is minimised by its nonlinearities and the sensitivity to temperature changes which is 50 to 100 times that for resistance strain gauges.

Transducer systems

Most of the transducers described above require the provision of a stabilised power supply and in many cases some special detection or amplifier circuit or phase sensitive rectifier. Where a system has a number of similar transistors, it may be possible to use a single power supply and possibly make some other economies in equipment. The continuing development of integrated circuits for operational amplifiers and other analog and also digital functions, has enabled many of these special circuits to be made sufficiently small to be built into the transducer itself. If signal conditioning and amplification is also included, the output of the transducer can be made more resistant to contamination and interference.

Index

absolute encoders, 14
absolute linear transducer, 92
accuracy of A/D conversion, 114
A/D converter, 113
address decoder, 7
address highway, 8
analog control, 4
analog conversion, 97
analog transducers, 9, 184
angular digital encoders, 13
aperture time, 114

ballscrew, 78
BCD, 178
binary codes, 178
brushes, 15
brushless synchros, 134

capacitive sensors, 167
capacitive tachometers, 49
capacitive transducers, 187
carrier variations, 159
coarse-fine synchros, 135
common mode, 111
computer controlled systems, 3
contact encoders, 14
control differential, 132
control transformer, 130
control transmitter, 130
conversion techniques, 118
converter errors, 159
counter, 54
crosstalk, 98
crystal oscillator, 57
cyclic codes, 17, 26

D/A converter, 5, 122
data highway, 8
data loggers, 1
density transducer, 70, 73
diaphragm, 63
digital filtering, 112
digital multiplexing, 10
digital ramp, 124
digital switch, 123
digital tachometers, 44
digital-to-synchro conversion, 162
displacement transducer, 78

earthing, 110
electromagnetic tachometer, 47
encoder lamps, 29
encoders
   linear, 80
   tachometer, 44
encoding disc, 13
errors, 159
excess Gray codes, 181

feedback A/C converter, 121
ferrostatic tachometer, 49
filtering
   digital, 112
   low pass, 109
flowmeter, 48
force balance, 168
frequency-dependent transducers, 51
frequency spectrum, 105
fringes, 87
function generator converter, 140

gas density transducers, 70
gas pressure transducers, 66
grating, 83, 87
grating/encoder system, 94
Gray code, 17, 180

harmonic-oscillator, 155
high resolution tracking converter, 148
highway system, 9
hysteresis in encoders, 28

incremental displacement transducer, 89
incremental encoder-linear, 83
incremental shaft encoders, 36
indexing systems, 25
inductance sensors, 165
inductosyn, 86
interference, 98
interpolator, 33, 90
interrogative transducers, 120

lamp life, 30
LDVT, 3
leadscrew, 80
LEDs in encoders, 30
level indicators, 164
linear encoders, 80
linear transducers, 78
liquid density transducers, 75
low level signal, 2
low pass filter, 105

magnetic encoder, 22
magnetic matrix, 171
magnetic shaft encoder, 21
mass flow measurement, 73
mechanical scanners, 99
Moiré fringe, 87
multiplexer, 2, 98
multitrack grating, 95

noise, 53, 98
noise error, 116
nonlinearity error, 116
normal mode, 110

offset error, 118
optical encoder, 24

optical isolator, 51
optical resolvers, 31

phase-shift converters, 137
phase-sensitive demodulation, 146
photocell amplifier, 27
position logic, 37
position sensors, 163
pressure transducer
   quartz crystal, 58
   vibrating cylinder, 66
pulse tachometers, 47

quadrant switching, 147
quadrature components, 146
quartz crystal, 57

radiation transducers, 173
recovery time, 110
reed switches, 100
reflecting light, 82, 87
relay multiplexer, 101
resistance strain gauges, 2, 184
resolver format, 136
resolvers, optical, 31, 35
resolver-synchro, 133
reticule, 81
rotary mechanical multiplexers, 100
rotor-to-stator errors, 159

sampling, 7
sampling converters, 151
sampling techniques, 3, 153
scale coil, 87
scanning head, 85, 88
scanning reticule, 81
scan problems, 16
Schmitt trigger, 27, 33, 39
sensors, photovoltaic, 26
signal classified, 11
signal conditioning, 2, 104
signal errors, 160
signal filtering, 109
straight binary code, 178
successive approximation, 125, 154
syncho incremental encoders, 40
synchro pair, 129
synchro pulse generator, 40

synchro/resolver errors, 159
synchro systems, 129

tachometers, digital, 44
temperature transducers, quartz crystal, 58
thermistor, 55
timer/counter, 54
timing and control, 3
toroidal pick-up, 22
tracking converters, 144
transistor multiplex, 103
two-level multiplexing, 6
two-speed synchros, 135
two-speed tracking converter, 149

U-scan, 20

V-disc, 19
vibrating beam transducers, 60
vibrating cylinder transducers, 64
vibrating diaphragm transducers, 63
vibrating string transducers, 59
vibrating tube transducers, 74
voltage/frequency converters, 51
volumetric flow, 76
vortex transducers, 176
V-scan 17

wear on brushes, 21